Computer Networking *for the* Small Business & Home Office

Other PROMPT® books by John Ross

Telecommunication Technologies
Guide to Satellite TV Technologies
DVD Player Fundamentals

Computer Networking *for the* Small Business & Home Office

by John Ross

ALEXANDRIA LIBRARY
ALEXANDRIA, VA 22304

©2001 by Sams Technical Publishing

PROMPT® Publications is an imprint of Sams Technical Publishing, 5436 W. 78th St., Indianapolis, IN 46268.

All rights reserved. No part of this book shall be reproduced, stored in a retrieval system, or transmitted by any means, electronic, mechanical, photocopying, recording, or otherwise, without written permission from the publisher. No patent liability is assumed with respect to the use of the information contained herein. While every precaution has been taken in the preparation of this book, the author, the publisher or seller assumes no responsibility for errors or omissions. Neither is any liability assumed for damages resulting from the use of information contained herein.

International Standard Book Number: 0-7906-1221-6
Library of Congress Catalog Card Number: 2001090724

Acquisitions Editor: Alice J. Tripp
Editor: Sara Black, BooksCraft, Inc.
Assistant Editors: Kim Heusel
Typesetting: Kim Heusel
Indexing: Kim Heusel
Cover Design: Christy Pierce
Graphics Conversion: Christy Pierce, Phil Velikan
Illustrations: Courtesy the author

Trademark Acknowledgments:
All product illustrations, product names and logos are trademarks of their respective manufacturers. All terms in this book that are known or suspected to be trademarks or services have been appropriately capitalized. PROMPT® Publications and Sams Technical Publishing cannot attest to the accuracy of this information. Use of an illustration, term or logo in this book should not be regarded as affecting the validity of any trademark or service mark.

PRINTED IN THE UNITED STATES OF AMERICA

9 8 7 6 5 4 3 2 1

Contents

Acknowledgments

I would like to thank the following companies and manufacturers for allowing me to use these images. The following companies and manufacturers do not state or imply an certification or approval of the material covered in this book.

Cisco Systems: Figures 1.1, 3.18a, 11.2, 11.5, 11.6, 11.7, 11.8, 11.9, 13.1, 13.2, 13.3, 13.4, 13.5, 13.6, 13.7

Fluke Networks: Figures 2.4, 2.5

Omni Systems Associates: Figure 2.6

AVI: Figure 3.11

National Semiconductor: Figures 3.18b, 11.18, 11.24

3Com Corporation: Figure 6.18, 11.10

Intel Corporation: Figures 7.4, 7.5

Hewlet-Packard: Figures 8.1, 8.2, 8.3, 8.4, 9.1, 9.2, 9.3, 9.4, 9.5

Citrix: Figure 10.1

Philips Semiconductors: Figures 11.22, 11.23

Home RF: Figures 11.26, 11.27, 11.28, 11.29

Quantum Corporation: Figure 12.2

Gadzoox Microsystems Inc.: Figures 12.3, 12.4, 12.5, 12.8

Preface

Computer Networking for the Small Business and Home Office takes a fresh approach to business networking. The book establishes a foundation for understanding network operation with discussions about linkages between the OSI model and network applications. In addition, it also takes the reader on a tour of networking trends such as Peer-to-Peer computing, distributed computing, Web servers, portals, and wireless networks that have the potential to take the world of small business networks to new heights.

The technical content provides a solid reference, but the book also supplies the information needed by managers and business owners to make solid decisions about networking. The book covers all aspects of small business network technologies using a writing style that makes the language of the technology easily understandable. As the chapter-by-chapter outline indicates, the book will take a step-by-step approach to the learning process.

The introduction to Chapter 1 establishes the purpose for the book and provides detailed information about the future of network technologies and applications, and the first four chapters lay the groundwork for the remainder of the text. Chapter 1 provides an overview of small business network fundamentals with a discussion about network design, topologies, and operating systems. The chapter discusses how computer networks function, defines different network types, and provides information needed for obtaining a clear understanding of networking.

Chapter 2 defines network management and provides information about software and hardware used for network management. The chapter categorizes network management requirements into the five functional areas of configuration management, fault management, security management, performance management, and accounting management. Within this framework, the chapter covers key software tools such as Remote Monitoring and Simple Network Management Protocol as well as hardware devices such as protocol analyzers. The chapter concludes with a discussion about capacity planning.

Chapters 3 and 4 use the OSI model for networking as a basis for discussions about the building blocks of networks. With coverage of the Physical and Data Link Layers, Chapter 3 also makes a connection between transmission standards and defines different transmission media used in local-area networks. The chapter briefly defines and compares the characteristics of twisted-pair cabling, coaxial cabling, and fiber-optic cabling. In addition, the chapter covers network interface cards and hubs.

As the book works through various types of network technologies, it continually emphasizes the relationships between the network technologies and the OSI model. Chapter 4 continues the discussion about the OSI model by considering the upper layers of the model. The chapter considers network layer devices routers, switches, and gateways, but it also covers protocols and applications. The layers, protocols, and reference models illustrated in chapter 4 are established as an organizing framework for the functionality and operation of a network.

Chapter 5 provides a tour of the server technologies and explains the processes used within the servers. In the opening stages of the tour, the chapter explains the differences between processors used within the servers and defines different server applications. From there, the chapter moves to a detailed examination of traditional server tasks as well as servers used for Web delivery, proxy access, printing, facsimile delivery, communications, and remote access. Chapter 5 concludes with a discussion of fault tolerance, server load balancing, and server clustering.

Chapter 6 describes applications for Ethernet, Fast Ethernet, and Gigabit Ethernet technologies. Given the widespread uses of Ethernet technologies, the chapter begins by describing how the Ethernet functions with respect to the OSI model. In addition, the chapter provides a detailed discussion about the Carrier Sense Multiple Access with Collision Detect protocol. The discussion in Chapter 5 moves from those key points to definitions and descriptions of applications for the Ethernet technologies. Within those discussions, the chapter also compares shared and switched networking.

Chapter 7 emphasizes Peer-to-Peer computing, client-server computing, and distributed computing. Even though the phrases "client-server" and "Peer-to-Peer" have become commonplace in network computing and usually describe some type of network function, the chapter breaks the phrases down into definitions of threads of execution. The chapter covers the potential Peer-to-Peer computing and describes different models of Peer-to-Peer communication.

Before discussing the different network architectures, chapter 7 defines the term "middleware" as a combination of architecture, programming language, communications program, a data manipulation program, programming interface, and translation driver. In turn, the chapter describes how this combination allows middleware to act as a bridge between systems and provides a method for allowing users to connect multiple data sources through networks. As the chapter moves to the overview of distributed computing, it shows that a Distributed Computing System interconnects autonomous computers through a communication network for the purpose of supporting business functions.

Chapter 8 complements Chapter 7 by showing that distributed computing applications take advantage of the flexibility given through objects, or reusable software components, and object-oriented programming. The chapter describes the performance of objects in terms of units and describes object-oriented programming. In a look back to Chapter 7, chapter 8 also shows how object technologies work as middleware tools. Middleware technologies and tools such as Application Programming Interfaces, Remote Procedure Calls, Interprocess Communication Tools, and message-based middleware interconnect data and processing resources across distributed networks.

Chapter 8 also covers object-brokering tools that extend the capabilities given through Remote Procedure Calls. The special-purpose middleware supports distributed object applications by allowing remotely located objects to communicate with one another. Examples of these tools include the Object Management Group's Commom Object Request Broker and Microsoft's ActiveX. The chapter concludes with an overview of application hosting and uses for the object-oriented technologies for that purpose.

With Chapter 9, the book shows how software applications called Enterprise Information Portals allow organizations to access stored information and provide users a single gateway to personalized information. With this capability, the portal establishes a format for "one-stop" information shopping and includes shared services such as security, metadata storage, and personalization. Chapter 9 also illustrates how the use of intelligent agents and software applications throughout a portal allows the management, analysis, and distribution of information within an enterprise and outside of the enterprise. In addition, the chapter describes how portals use intelligent agents to provide the capability to interact with customers and support the bidirectional exchange of information.

Chapter 10 portrays thin clients as the next step in the computer and networking technology evolution. The chapter describes methods used in thin client networks to establish information access and delivery rather than information processing. In addition, the chapter considers the use of network servers to provide the processing power for the thin client network and the role of the client machines.

In Chapter 11, the book shows that wireless local-area networks can augment rather than replace wired local-area networks by providing the final few meters of connectivity between a backbone network and the in-building or mobile user. In addition, the chapter illustrates the use of wireless technologies to reduce the cost of network ownership through portability and scalability. The chapter provides detailed information about the wireless local-area and personal-area network standards and protocols while describing the differences between the IEEE 802.11, Bluetooth, and HomeRF standards. Chapter 11 also defines the hardware used with wireless networks.

Chapter 12 addresses Storage Area Networking applications and issues. Most major networking vendors agree that Storage Area Networks will assume a major role in distributed computing and data warehouse environments. Chapter 12 builds from those opinions to show methods for attaching and sharing devices within a Storage Area Network. The chapter describes current uses for storage area networks including the connection of shared storage arrays, the use of clustered servers in a server failover environment, the utilization of tape resources to network servers and clients, and the creation of parallel and alternate data paths for high performance or high availability computer environments. Chapter 12 concludes with a discussion of Network Attached Storage.

Chapter 13 compares uses of Virtual Private Networks and Virtual Local-Area Networks. Within this chapter, the reader finds how a Virtual Private Network uses the Internet or other network service as its wide-area network backbone. In addition, the chapter shows that Virtual Private Networks use local connections to an Internet Service Provider or the point of presence of other service providers to replace dial-up connections to remote users and leased-line or Frame Relay connections to remote sites. From there, the chapter describes how a

Virtual Private Network extends a private Intranet across the Internet or other network service and facilitates secure eCommerce and Extranet connections between business partners, suppliers, and customers. Chapter 13 continues with a definition of virtual local-area networks and methods used for using the Virtual Local Area Network to create the effect of a local-area network over a wide-area network.

Chapter 14 gives the reader a glimpse of the security challenges facing network technicians. Because an organization defines a set of access rights, privileges, and authorizations and assigns those to appropriate individuals working under the organization, a major portion of Chapter 14 discusses policy-making decisions. The chapter uses those discussions to frame other discussions about security policies and methods. With this, the chapter covers viruses and virus protection software, e-mail user standards, and user privileges along with authentication and access control practices.

Chapter 14 also defines authentication tools and policies as system-wide tools for recognizing and verifying the identity of users. In addition, the chapter considers access control methods used either to provide or to reject access to some service or data in a system. As the chapter concludes, it considers security measures applied at the IP layer of the Internet and the use of firewalls to protect organizational information.

Dedication

This book is dedicated to my parents, John C. and Lorraine Ross, and all my friends at Fort Hays State University. My parents gave the love and support that one needs when working a large and difficult project. My mother also provided assistance with the artwork that complements the text. Many times she worked the same late-night hours as I so that the deadlines would be manageable. At the university, I've received support and congratulations from the faculty, staff, and students. The staff at Forsyth Library has been wonderful in providing assistance as I attempted to locate research materials. I also thank President Edward Hammond, Provost Lawrence Gould, and Vice President Bruce Shubert for supporting me along the way and for challenging me to achieve excellence.

1

Small Business Network Fundamentals

Introduction

Computer networks are commonplace in organizations that rely on communication between multiple computer systems, peripherals, mass storage systems, and other devices. Networks also allow users to exchange electronic mail, share licensed applications software, gain access to common resources, and collaborate on projects. Given this flexibility, the number of small-business network installations has increased dramatically.

In addition to flexibility, the capabilities of networks have also triggered growth. Compared to early designs, modern networks offer faster data transmission speeds and greater bandwidth. Architectures have evolved from simple Peer-to-

Peer network designs to client-server and thin client designs and then to hybrid Peer-to-Peer designs that utilize the Internet. The convergence of voice, video, and data communications technologies has produced desktop videoconferencing, video on-demand, virtual private networks, and voice over IP solutions for business tasks.

Networking Essentials

Networks provide the lowest level of information transport in client-server and distributed computing environments. From this perspective, a network consists of equipment and physical media that interconnect two or more computers. Because a wide variety of network configurations exists, the

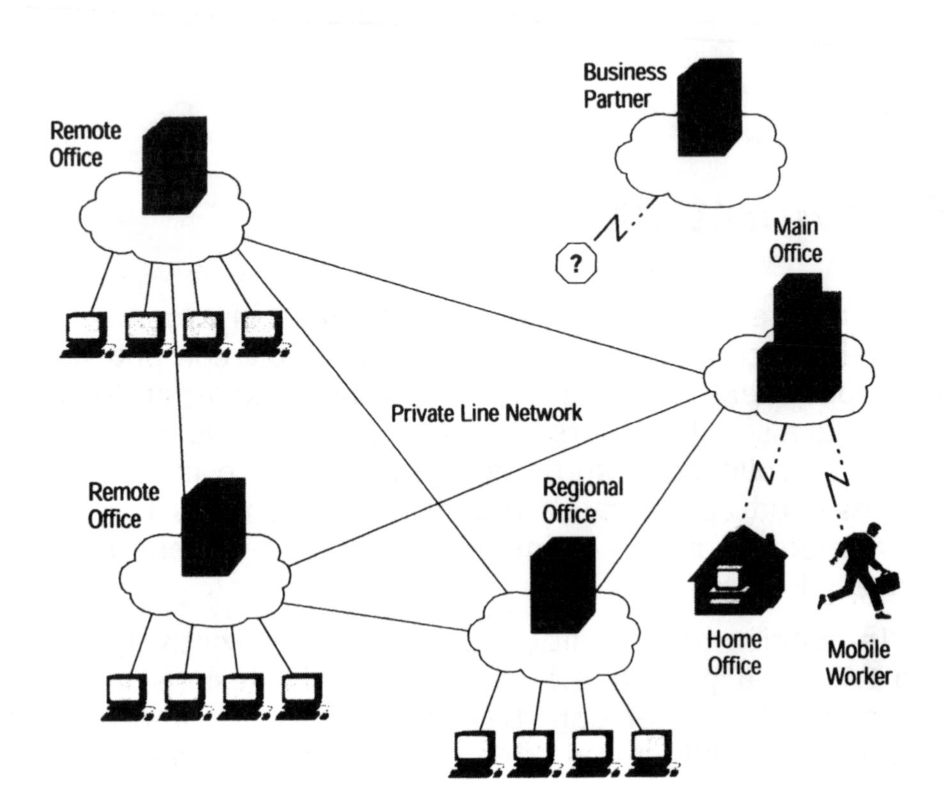

Figure 1.1. Modern corporate network.

size of a network may range from three desktop computers connected through a cable to a large group of computer systems that serve an international airlines reservation system. The latter system could utilize global communication satellites, large processors, and thousands of terminals and workstations. Local-area networks, or LANs, connect electronic devices within a single building or a cluster of adjacent buildings. Most local-area networks operate within a radius of six miles. Remote bridging allows LANs to extend beyond the traditional limits.

Pictured in figure 1.1, modern local-area network designs must provide the capability to connect across proprietary systems and form a single system with comparable physical interfaces and control protocols. To ensure flexibility for the system, the design must not only satisfy current needs for the transmission of voice, video, and data information but also accommodate technological advances. Because users require the capability to communicate with other users located outside the system, the LAN also must serve as an interface or gateway for that communication.

Network Development Strategies

Most network design teams employ either the top-down, bottom-up, or middle-out strategy for designing and deploying local-area networks. When choosing the correct strategy, a network designer will consider the size of the organization, the depth of existing in-house resources to guide the development and implementation of the network, and the existing relationships with information system vendors. Each of these factors defines the choices for cabling, hardware, network operating system, and applications software, which will become apparent throughout the network design cycle shown on the following page in figure 1.2.

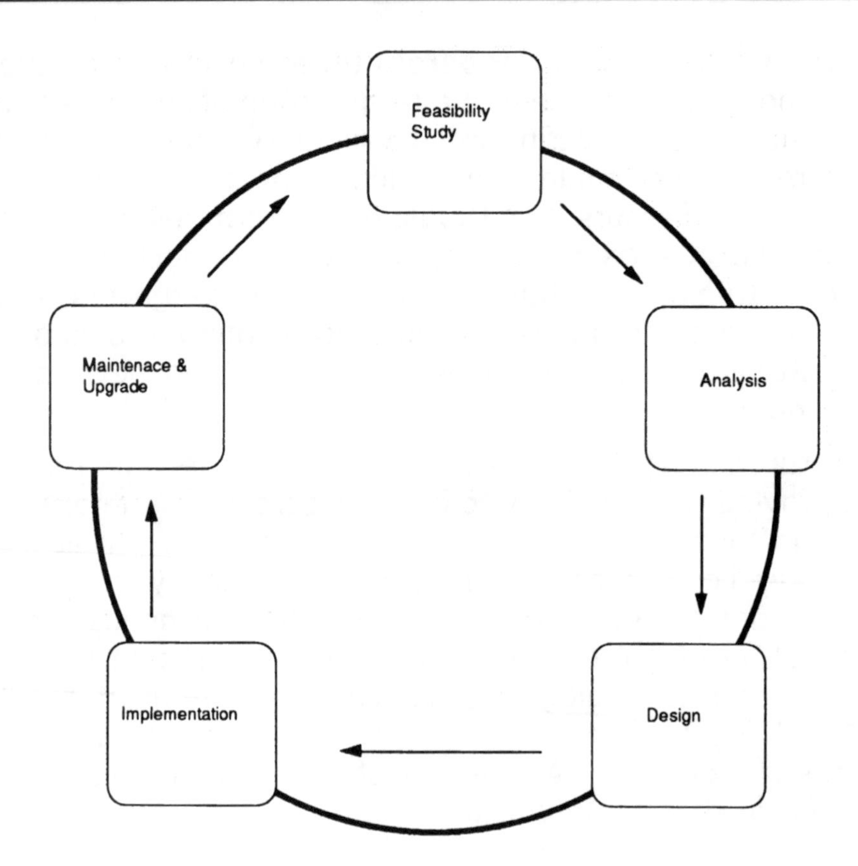

Figure 1.2. Network design cycle

Top-Down Development

When an organization deploys a local-area network as a strategic tool for enterprise computing, the network designer will usually manage the development process in a top-down method. At the forefront of the process, the organization establishes a single networking standard high within the information management hierarchy. As a result, longtime vendor relationships play a larger role in setting the course of action.

The top-down strategy recognizes the need to maintain compatibility with existing systems or a vendor's overall strategy. In these cases, adherence to the overall enterprise strategy rather than the applications intended for the network becomes the major emphasis. If the use of the top-down strategy is not carefully defined and employed, the type of real work accomplished on a network can be distorted and the fact that smaller units within the organization may utilize the network as a strategic tool may be ignored.

Bottom-Up Development

Bottom-up development of a network design usually occurs when the network designer knows more about technical components than the needs of the individual network users. Decisions solely based on technical expertise typically result in networks that answer current needs but fail to consider a wider range of options or future choices. Designers who base the usefulness of a network on the ease of installation and compatibility of the hardware and software often choose a topology that they understand. However, they lose sight of the fact that all subsequent use of the network also depends on that topology.

Middle-Out Development

The middle-out approach to network development depends on thoroughly analyzing the application of the network at the user level. In brief, this approach begins with the observation of all continuing tasks that use the resources given by the network. In addition, the middle-out approach focuses on current and long-term application needs. Because of these two factors, developers have the opportunity to study the managerial and authority patterns at the implementation site and to determine the capabilities of users. In turn, the middle-out approach discloses the amount of effort needed to support the network and provides an opportunity to survey the physical layout of the network from a different perspective.

Policies, Practices, and Procedures

Network managers and system administrators work with policies, practices, and procedures. At the most general and abstract level, policies describe rules for network users. As an example, a policy might show that every network user has access to e-mail or that an organization will maintain confidentiality for all users. Most organizations express policies in ordinary business language and do not use policies to address implementation methods.

The most common subjects for policies include the distribution of resources, the definition of security requirements, and the setting of performance expectations. Resource distribution includes defining accounts and access permissions, determining desktop configurations, outlining roles for mobile computing, establishing available applications, and setting user responsibilities. Security policies include procedures for physically securing the premises and for backing up and restoring data, plans for disaster recovery, and network security items such as data integrity, authentication, access control, virus protection, encryption, and intrusion detection. Performance policies set priorities among users, groups, and applications.

At the next level, practices narrow the scope of the policy and address implementation methods but not specific products or data structures. With this approach, a practice may take the first policy statement about all employees having access to electronic mail and show that all new employees will receive an e-mail account during the first work day and a schedule for training. Practices provide the advantage of establishing vendor- or product-independent methods of operation.

A procedure narrows the focus of the statement even more. Going back to the first example, a procedure regarding electronic mail could specify a naming system for e-mail accounts such as Firstname_Lastname and show that employees with the least seniority and duplicate names will use Firstname_Middlename_Lastname. Furthermore, a procedure about e-mail usage may address the responsibility of archiving messages or changing passwords.

Network Architecture

The phrase "network architecture" describes the combination of physical components, the functions performed by the components, and the interfaces between the components of a network. Transmission media consists of the cables or radio frequencies that carry information throughout the local-area network. Components and interfaces found within the network include extenders, hubs, switches, routers, and bridges.

Networks transmit signals using either the broadband or baseband technique. Broadband communication divides the transmission media into many channels and assigns a specific frequency to each channel. Similar to the technology used to broadcast television signals, frequency division multiplexing enables the same cable to carry many different signals at the same time and results in a high aggregate transmission speed. Broadband transmission requires the use of frequency translators and uses radio frequency modulation.

Baseband communication uses the entire cable as a single communication channel. Because baseband communication relies on digital technologies, it does not require a modem at either end of the cable. The use of a single communications channel requires the control of traffic throughout the network. Protocols provide this type of control and allow several users to share a single cable.

Network Standards and Services

Network standards ensure that compatibility exists between networks manufactured by different vendors and that the interconnection of those networks will occur without problems. For example, a university library that relies on a Sun-supported network may need to download information from another library that uses an IBM-supplied network. Network services provide the basic addressing and transport mechanisms needed to operate the network. The services communicate with the server and client middleware and include common and proprietary protocols. With the connectivity provided through services, network connectivity can occur within a given location or between remote sites.

Network Topologies

The topology of a network describes the design of the network. Network topologies consider interoperability, or the compatibility of resources on the network, and integration, or how hardware, software, and other resources work together. With all this responsibility, the topology establishes a standardized approach to accessing information and increases the value of the technologies.

Selecting a Topology

Understanding the size of any network becomes essential to determining the value of one network topology over the other. In turn, an understanding of the topology of LAN technologies can assist with planning for the LAN and provide alternatives for the installation or expansion of the network. At the most basic level, the topology of a network refers to the method used to connect all the parts of a network. A network topology creates a layout of the computers, printers, and other equipment connected to the network.

The choice of an appropriate LAN topology depends on the transmission component and the control mechanism. In turn, these factors dictate the physical topology of the network. In some cases, however, the physical characteristics of a site determine the topology chosen for the network. In other cases, a decision about the network operating system affects decisions about the topology.

A decision to build the network upward from the physical level also determines the type of transmission and topology employed for the network. After making that decision, a network manager can select an appropriate access method, interface, and operating system. In contrast to building upward, a network manager may decide to build downward from the application level. With this scheme, the manager selects the application software and network operating system before making a decision about the access method and topology.

Because cable may connect resources into a network, the network topology also depicts the organization of the cabling. The three basic physical topologies available to network designers include the bus, the ring, and the star. Although recent technological advances have blurred the distinctions between the physical and logical arrangements, the selected topology may also dictate the use of a specific media-access control method such as Ethernet or Token Ring for the network operation.

Because the logical topology defines an electrical path and the physical topology defines the arrangement of cables, connectors, and nodes, the logical layout of a network may differ from the physical layout. As an example, Ethernet operates as a logical bus network but may have a physical star or bus topology. A Token Ring network operates as a logical ring but configures as a physical star topology.

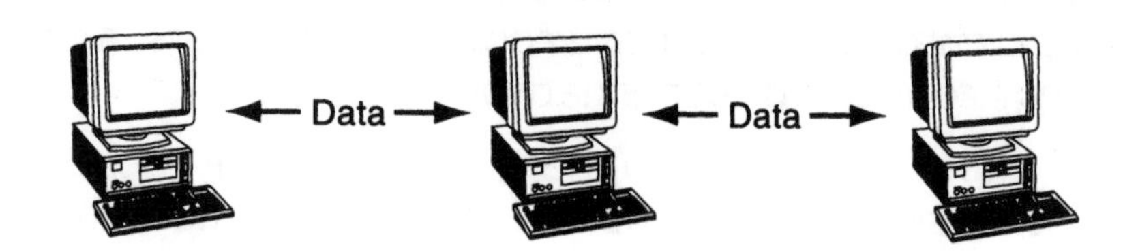

Figure 1.3. Diagram of bus topology

Bus Topology

Pictured in figure 1.3, the bus topology attaches all workstations on the network to a bus, or a linear, bidirectional communication path with defined end points. At either end of the bus, a terminator provides the resistance needed to establish the load for attached devices. A signal transmitted from a node proceeds along the bus in both directions. As signals pass by, each node listens for its specific address. When the node recognizes an address, it accepts the signals. Otherwise, the node ignores the data.

The bus topology controls access to the network through either centralized or distributed methods. Distributed control allows sharing by all nodes, and centralized control relies on a central controller. Although the bus topology uses cable more efficiently than other configurations and has a lower installation cost, it requires a bidirectional control system to prevent data collisions. Breaks in the bus affect only the nodes attached after the break. Ethernet and AppleTalk networks operate as bus-based networks.

Star Topology

Figure 1.4 shows that each node of the star topology connects directly to a central wiring controller such as a hub or switch through an individual length of cable. The cabled nodes seem to radiate from the controller; however, each connection offers bidirectional capabilities. As a result, the star topology offers the advantage of centralizing key networking resources and gives the network administrator a focal point for network management.

Because a network based on the star topology features each workstation connected to a central device through a dedicated cable, the topology requires a larger investment in cable. However, the topology offers advantages through the easy modification of cables. In addition, a network administrator can easily detect and isolate network failures.

Transmissions pass from the transmitting node to the controller and from the controller to the receiving node. The controller manages and controls all communication. The star topology relies on the sending of packets from one station to another with the packets repeated to all ports on the hub. The sending and repeating of the data packets allows all stations to see each packet sent on the network. However, only the station with the appropriate packet address receives the data.

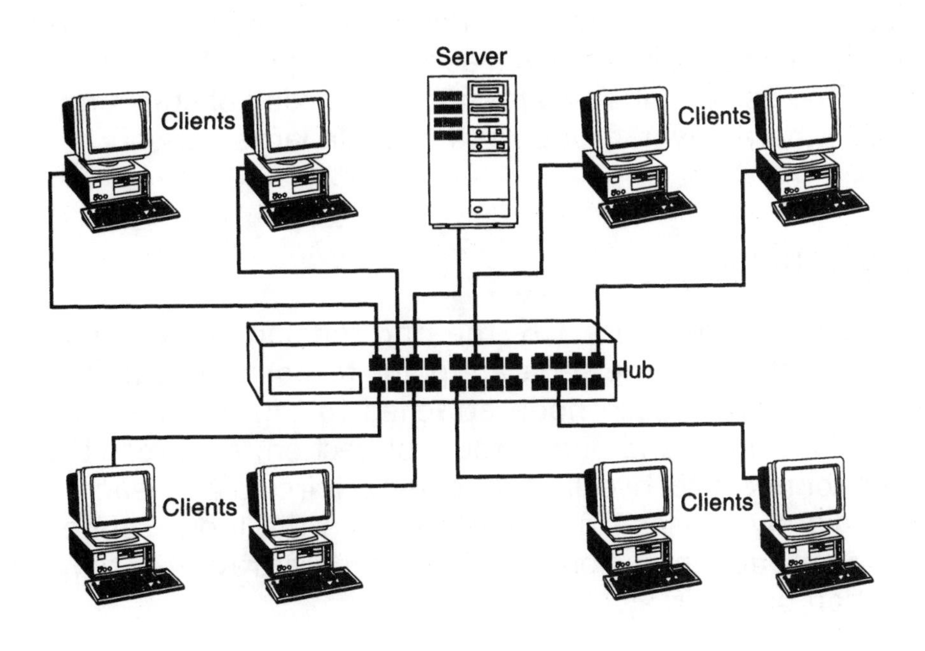

Figure 1.4. Star topology

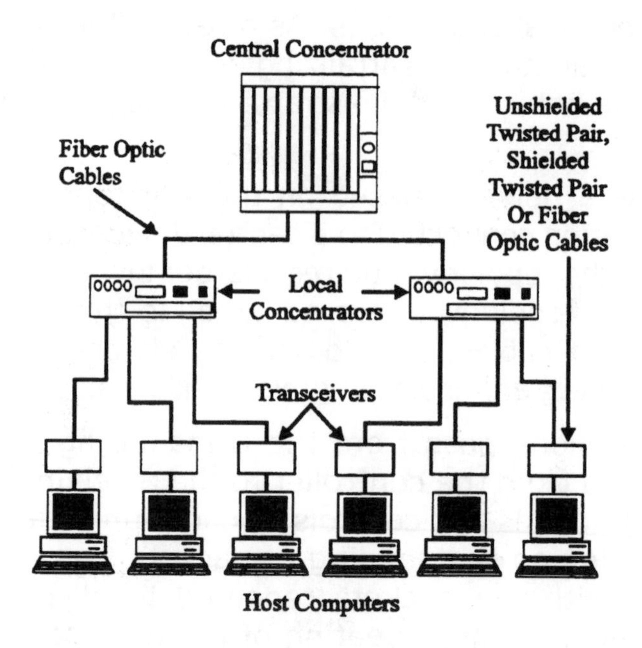

Figure 1.5. Ethernet network

In figure 1.5, a hub with Ethernet is used to allow the creation of larger networks that consist of cascaded stars. In this arrangement, one hub serves as the focal point for many other hubs. The use of hubs also permits the mixing of star- and bus-based Ethernet workgroups into a single large network.

Ring Topology

As shown in figure 1.6, the ring topology takes its name from the shape of the separate point-to-point links that make up the network. Each node attached to the ring network includes one input and one output connection. As a result, each node connects to two links. Repeater circuitry in each node immediately passes signals received on the input connection to the output connection. As a result, data travels in only one direction on the ring.

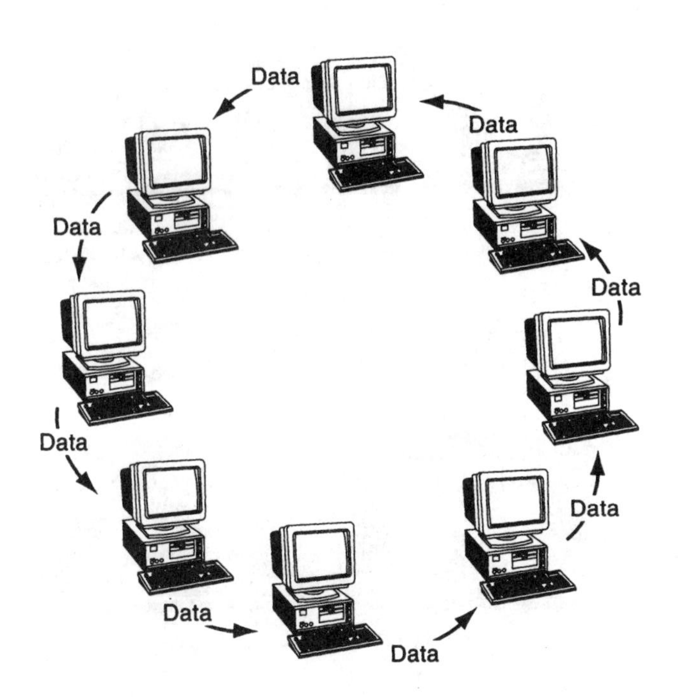

Figure 1.6. Diagram of ring toplolgy

Networks based on the ring topology became popular when IBM introduced its Token Ring technology in local-area networking. Like the bus topology, the ring topology relies on a single cable. However, the cables in a ring topology loop and form a complete logical circle or ring. Each workstation connects directly to a central device called a Media Access Unit, or MAU.

Unlike the bus topology, the ring topology uses a deterministic — rather than a contention-based — access method. With each node in an active state, all nodes have the ability to place new messages on the ring. As the messages circle the ring, each has an address for a specific node that allows the node to copy the message. If a node fails, then the failed node does not repeat signals from the input. The failure of any node breaks the ring and causes the transfer of data to stop.

With Token Ring networks similar to that shown in figure 1.7, an electronic signal called a token passes from station to station on the ring. Each station regenerates the token as it

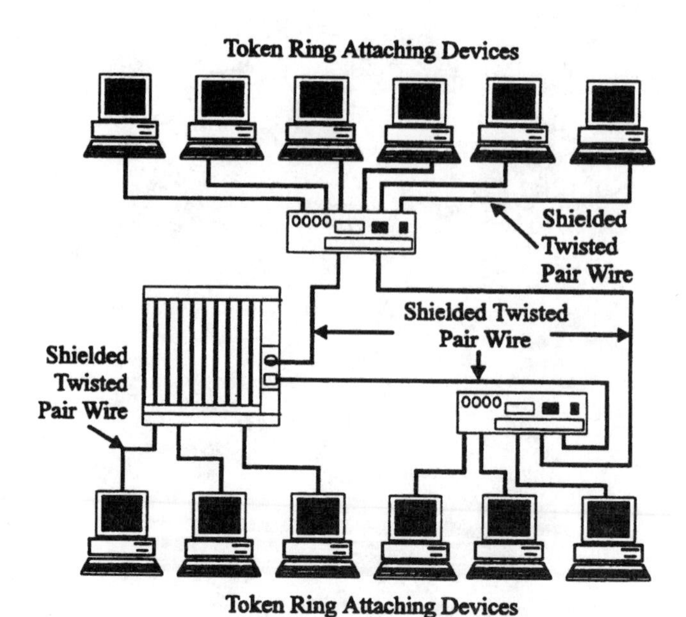

Figure 1.7. Token Ring network

passes by the node. When a station wishes to transmit data over the network, it must wait until the neighboring station passes the token. As soon as the station takes control of the token, it can place a data packet on the network. After the data packet makes a full circuit of the ring and returns to the originating station, the receiving station releases the token for the next workstation.

Mesh Topology

The mesh topology connects every node on the network to every other node on the network through many different paths. Because of the large number of connections, the mesh topology has not gained widespread use in local-area networks. Figure 1.8 shows the diagram for the mesh topology.

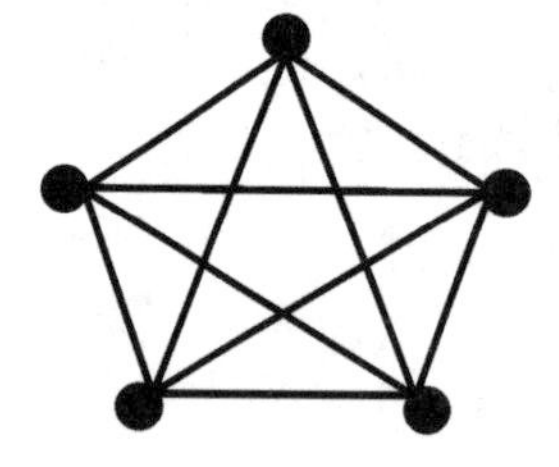

Figure 1.8. Diagram of mesh topology

Tree Topology

The tree topology operates as an expanded bus configuration. As figure 1.9 indicates, the tree topology turns the bus on end. Individual nodes of the tree topology extend into branches. Because the tree topology builds off the bus topology, it offers the same advantages and disadvantages seen with the bus.

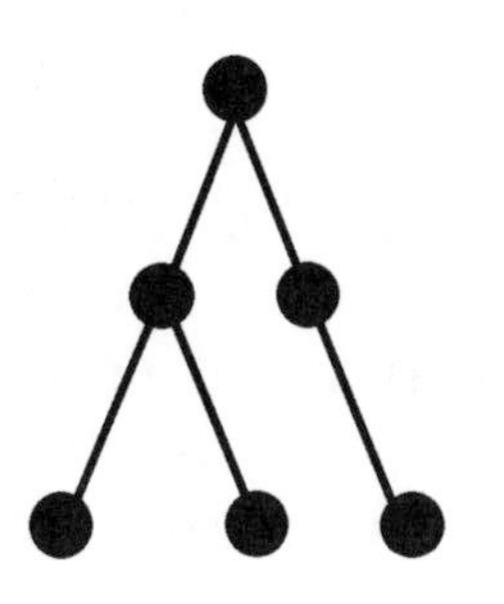

Figure 1.9. Diagram of tree topology

Hybrid Topologies

Referring to figure 1.10, a hybrid topology contains elements of more than one network configuration. As the figure shows, a bus network may have a ring network as a link. Another type of hybrid network features a star network that has a bus network as a link.

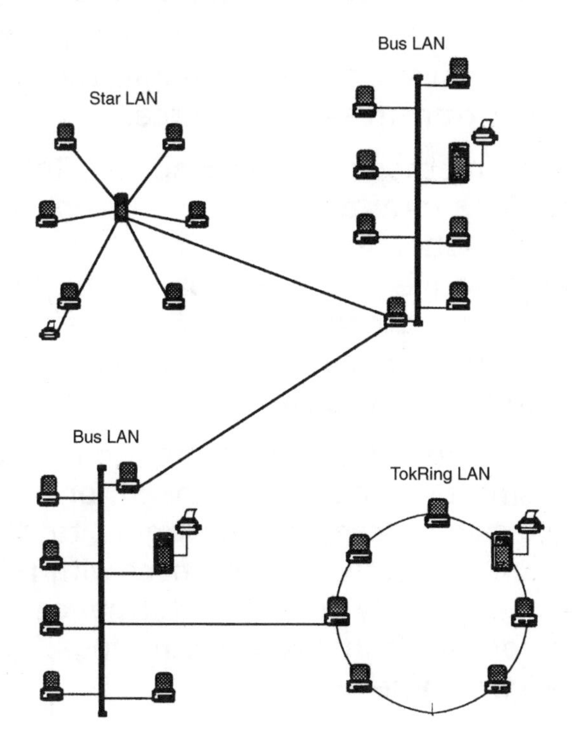

Figure 1.10. Diagram of hybrid topology

Access Methods

Access methods carry out network processes that guarantee the error-free delivery of information between systems. Given the importance of the overall design of a LAN, the decision about access methods requires attention to factors such as cost, reliability, and compatibility. An access method has the greatest impact on the performance of a local-area network because it defines specific components that affect network speed and reliability. Mismatching access methods and system requirements will cause performance problems as well as higher costs per connection.

In brief, access methods for local-area networks may occur in either a centralized or a distributed environment. Most LANs utilize distributed access methods that have each workstation participating equally in the control of the network. In the contention scheme of distributed access, any station can initiate a transmission at any time. In the deterministic access method, each station must wait for a designated time before transmitting. Deterministic access manages the sequential allocation of network transmission resources.

Popular access methods include Ethernet, Token Ring, the Fiber Distributed Data Interface, and wireless options. Ethernet provides the best solution for networks that include many users who generate relatively little network traffic over long time spans. The contention method used for Ethernet networks does not operate well under sustained, heavy traffic loads because high contention levels slow network performance.

A Token Ring network avoids contention by distributing the right to transmit by circulating a token, or a distinct data bit pattern that assigns transmission rights to the receiving workstation. Token Ring sites operate best with relatively few users and provide relatively steady performance under heavy traffic loads. Performance in large Token Rings, however, degrades under heavy traffic loads.

The fiber distributed data interface, or FDDI, delivers the high speed and high bandwidth made available through the use of fiber optic cable. Even though Ethernet networks have higher usage rates, FDDI continues to remain in place because of the capacity to achieve high speeds for graphics, voice, and video applications through fiber optic connections. Wireless options such as the 802.11 and Bluetooth standards use infrared, radio transmission, or cellular phone technologies while mimicking the topologies seen with wired networks.

Network Traffic

Batch traffic consists of large transfers of data packets in one direction and occurs in both Peer-to-Peer and client-server applications. As an example, batch traffic from a server involves such services as server-based file and printer redirection, server-resident applications, and the transfer of files from one destination to another. Because of the transfer of large amounts of data, batch traffic requires greater bandwidth.

In comparison to batch traffic, interactive traffic consists of repeated, small data transfers to and from a server. Because the quantity of data per packet remains at less than 100 bytes, interactive traffic offers high transaction rates, high frame rates, and low bandwidth usage. In most instances, interactive traffic involves mission-critical applications such as on-line transaction processing, credit inquiries, or fund transfers. With each, simultaneous interactive transactions occur between a server holding a large database and numerous clients.

Client-server traffic strikes a balance between batch and interactive traffic on several levels. As with interactive traffic, client-server traffic uses a high-performance server and involves bidirectional data transactions. In contrast with interactive traffic and like batch traffic, client-server traffic involves large data transactions and requires more bandwidth. Client-server applications include database transactions, electronic mail, and financial transactions.

Understanding the Bandwidth

Requirements of each transaction requires an understanding of the total traffic flow — or the size of the transaction — and the duration of each transaction. Interactive transactions produce relatively little bandwidth; however, high-performance client-server traffic can deliver throughput in the single-digit megabits-per-second (Mbps) range. Batch traffic can consume several megabits per second per connection. As an example, a single network file transfer accomplished through the File Transfer Protocol can use more than 15 Mbps of bandwidth.

In addition to appreciating traffic size and duration, an understanding of network usage — or how users operate on the network — also becomes useful for predicting network requirements. As an example, a network of 100 users that has transactions occurring at a rate of two transactions every 10 seconds has the same amount of traffic seen with a network of 200 users that has transactions occurring at a rate of one transaction every 10 seconds. By calculating the number of users on the network and the percentage of time that each user remains active, a network designer or systems administrator can estimate how often a certain number will remain active at the same time. In a two-server network that has 50 clients where an average of 5 percent of the clients remain active at any given time, over 99 percent of the time less than eight clients will have concurrent active operations.

Network Transmission Methods

Local-area networks may use unicast, multicast, or broadcast data transmission methods. With each type of transmission, the transmitting node sends a single packet to one or more nodes. In a unicast transmission, the transmitting node sends a single packet to a destination on a network. Prior to transmission, the source node addresses the packet by using the address of the destination node. From there, the transmitting node sends the package onto the network where the packet passes to its destination.

With multicast transmission, the network copies a single data packet and then sends the packet to a specific subset of network nodes. As with unicast transmission, multicast transmission addresses the packet prior to sending. However, the source node addresses the packet by using a multicast address. As soon as addressing occurs, the transmitting node sends the packet into the network. Then, the network copies the packet and sends a copy to each node that is part of the multicast address.

A broadcast transmission works much like the multicast transmission. A key difference exists, though, in that the network copies the single data packet to all nodes on the network. The source node addresses the packet by using a broadcast network.

Network Operating Systems

LANtastic

LANtastic 8.0 provides seamless connectivity to different computing environments by connecting workstations operating with Windows NT 4.0, Windows 95/98, Windows 3.X, or DOS into one network. As a result, an organization can add personal computers to the network without upgrading the operating system. Along with proprietary Applications Programming Interface support, LANtastic supports the TCP/IP, Ethernet, Token Ring, and NetBIOS protocols. The network operating software offers security methods such as Agent Communication Language group security, password protection, time-restricted user logons, and password expiration. LAN management features include audit trails, diagnostic indicators, remote accounts, and remote server management.

LANtastic does not require a dedicated server. LAN services provided through LANtastic include file and printer sharing, print spooling, and CD-ROM sharing. An integrated groupware system includes advanced e-mailing, faxing, paging, and scheduling features. In addition, network users connected through LANtastic can communicate with one another through a chat feature.

Banyan VINES

Banyan Virtual Integrated Network Service, or VINES, implements a distributed network-operating system based on a proprietary protocol family. With the protocols derived from the Xerox Corporation's Xerox Network Systems, or XNS protocols, VINES uses a client-server architecture where clients request certain services such as file and printer access from servers. In the VINES network scheme, all servers with multiple interfaces have an operation similar to that seen with routers. At the heart of VINES, a file system called the VINES File Store, or VFS, addresses file-sharing issues such as access rights, directory synchronization, and file reconciliation for heterogeneous networks. Network administrators can set rights at the file level; VINES matches security rights across platforms.

The Banyan network operating system relies on the VINES Internetwork Protocol, or VIP, to perform internetwork routing. In addition, VINES also supports a proprietary version of the Address-Resolution Protocol, a proprietary version of the Routing Information Protocol, or RIP, called the Routing Table Protocol, or RTP, and the Internet Control Protocol, or ICP. In addition to featuring proprietary protocols, VINES also supports the TCP/IP, NetBIOS, SNA, X.25, Token Ring, and Ethernet protocols. The software includes AppleTalk Filing Protocol for support for Macintosh workstations and SMB extended support for support of OS/2 workstations.

Apple Networking Products and the Macintosh Operating System

Apple Computer offers a series of networking products that operate in different environments. AppleTalk covers three communications protocols that change data transfer speeds depending on whether the user has access to an Ethernet-based network, a token ring-based network, or a Macintosh network. EtherTalk allows 10 Mbps access to Ethernet networks, TokenTalk provides 4-Mbps access to token ring networks. LocalTalk permits Macintosh computers to communicate through any network configuration at 230.4 kilobits per second (kbps).

Apple's Macintosh operating system evolved from the first successful graphical user interface introduced during 1984. The system does not run on Intel x86 or Pentium microprocessor architecture and requires either Motorola 689x9 CISC- or PowerPC RISC-based computers. At startup, the core Macintosh operating system and a series of user-controlled options, patches, and feature-enhancing plug-ins load into the random-access memory from the local hard disk and the read-only memory. All Macintosh files have an icon and can have names up to 31 characters in length. File management occurs through the dragging and dropping of files within the same volume, to other volumes, or across a network.

The Macintosh operating system provides fast and user-friendly networking through its Open Transport networking software. With the inclusion of Open Transport PPP, a control panel provides for the monitoring of dial-up TCP/IP Internet connections. The system also supports multiple Internet configuration settings without the need for rebooting and stores World Wide Web, FTP, and Gopher sites as Finder icons. The Apple Remote Access, or ARA, client allows Macintosh computers to call into Macintosh networks across normal phone lines.

Another feature called the Apple Internet Connection Kit, or AICK, provides automatic configuration and registration for the setup of an Internet connection. The implementation of Cyberdog allows users to save and launch URLs directly from the finder, embed live Internet links into documents, and create Internet-live cyber-documents. A user can store Internet URLs, network server names, local and network files, and electronic mail addresses in one location. Cyberdog includes the capability to perform a full-text indexed search for e-mail and allows multiple e-mail accounts.

The Macintosh operating system also includes built-in decompression and viewers for HTML, FTP, and Gopher. Users can remotely manage an FTP server by dragging files, renaming folders, and rearranging files and folders. Apple's QuickTime Media Layer, or QTML, enables cross platform authoring and playback with elements such as video, music, virtual reality, three-dimensional graphics, text-to-speech conversion, and teleconferencing. The OpenDoc Essentials Kit incorporates an image viewer that allows the viewing of PICT, GIF, JPEG, and TIFF files.

Linux

Linux operates as a freely available clone of the UNIX operating system and relies on commands similar to those found in the UNIX command set. In addition to using very little memory, the operating system supports a full range of system and network administration needs. Third-party software development tools allow the importing of code from other operating systems and the support of applications software.

Novell NetWare

Offered by Novell, NetWare provides a distributed, multitasking local-area network operating system. The latest versions of NetWare feature improved NetWare Directory Services and TCP/IP support that facilitates simpler installation and administration. The NetWare Directory Services, or NDS, maintains information about every resource on the network and provides network administrators and users with a consistent view of the network.

NetWare provides integrated administrative features that ensure data integrity when network failures occur. Although all LAN software provides access protection on files and locking facilities at the record level, NetWare provides disk duplexing, data mirroring, and transactional control of backout and information recovery. The central administration system contained within NetWare stores user data on disk and notifies users about their rights and privileges as they login into the network.

NetWare Directory Services

NDS operates as a distributed database and can replicate and partition databases for improved performance and fault tolerance. An NDS object consists of information divisions called properties and the data contained within the properties. A property refers to the categories of information that a user stores in the database for NDS objects. NDS management involves creating and controlling objects. Each object has properties containing information about that object. When the network requests information from the NDS, the directory searches each user object in the database and returns a list of all the objects that have the requested information within the respective properties.

The NetWare Directory Tree represents a logical operating structure for the NDS. Established as a hierarchical tree structure, the Directory Tree starts at the root object, branches out, and consists of container objects and leaf objects. The container objects act like directories in a file system and group information, and the leaf objects represent network resources such as computers, printers, lists, and users. A branch of the Directory Tree consists of a container object and the objects held by the container object. Leaf objects set at the end of the branches and do not contain other objects.

NetWare Web Server

The NetWare Web Server gives customers a method for using the Internet for both internal and external publishing. Users can take advantage of open standards to access public networks and private Intranets. In addition, customers may use the Internet to link geographically dispersed LANs, provide LAN-to-LAN file sharing, and enable the management of remote LANs without privately managed network connections.

NetWare Client Support

Novell offers a family of 32-bit client software that provides full 32-bit access to NetWare services including access to NDS and simultaneous access to multiple network servers, printers, and application resources. Three different versions of Netware Client 32 support Windows 3.1, Windows 3.11, Windows 95/98, and Windows NT. NetWare Client 32 for Windows 95 offers seamless integration of Windows 95 systems to NetWare services and includes the NetWare Application Manager that allows administrators to manage and access desktop and network applications across the network through NDS.

Netware Client for Windows NT connects Microsoft Windows NT users to all NetWare services. Administrators can use the client software to centrally manage NetWare services directly from Windows NT workstations and to dynamically refresh client software across the network. The Application Launcher utilizes NDS to give users location-independent access to their applications. As a result, network managers can centralize application installation and administration. Users may also dial out to remote NetWare or NT servers.

Microsoft Windows

Microsoft Windows 95/98

Introduced during 1995, Windows 95 offered improved performance and data integrity when compared to previous versions of the Windows operating system. Windows 95 implemented a 32-bit preemptively multitasking and multithreading kernel, improved resource usage and crash protection, and enabled the use of filenames that have a maximum length of 255 characters. In addition, Windows 95 introduced the Win32-bit Applications Programming Interface, or API.

Windows 95 also introduced a more complete package of network clients, protocols, and administration and resource-sharing components than was seen with previous versions of Windows. Because Windows identifies some networking components as protected mode network components, support features for Novell NetWare LAN client, Windows NT LAN client, remote network access, and file sharing run in extended memory. In addition, the Windows 95 support of real-mode network drivers and clients enabled Windows 95-based computers to run on non-Novell and non-Microsoft networks. The introduction of Windows 95 also integrated the TCP/IP protocol stack into the Windows operating system and allowed Windows 95 to communicate with the Internet.

As an upgrade to Windows 95, Windows 98 aggregates three years of various software patches and includes support for hardware produced after early 1995. Enhancements found within Windows 98 also include the integration of a set of Internet tools including:

- An active desktop and browser-based shell.
- Support for Internet standards.
- An improved mailbox client.
- Support for Internet conferencing.
- Support for streaming multimedia.
- Support for Web site creation and management tools.
- Improvements to the dial-up networking module.

Of particular note, the implementation of the Active Desktop and Browser-based shell allows the system to treat the Internet as an extension of the Windows file system. As a result, Windows Explorer can browse the local disk, a network drive, or the Internet. The improved mail client supports POP3/SMTP e-mail, and the dial-up networking module supports ISDN, Multilink Channel Aggregation, and Virtual Private Networks. In terms of multimedia support, Windows 98 enables the playback of MPEG audio and video files, AVI video, and QuickTime video with its ActiveMovie video streaming architecture.

Networking support integrated within Windows 98 includes support for the Distributed Component Object Model, or DCOM, which provides the infrastructure needed to support cross-network applications. The Microsoft package also includes a TCP/IP protocol stack and the NetBeui and IPX protocols. In addition, Windows 98 offers compatibility with Novell NetWare through support for the Netware Directory Services. Windows 98 also includes drivers for hardware classes such as the Universal Serial Bus, IEEE 1394 FireWire, and IrDA infrared communications.

Microsoft Windows NT

The Microsoft Windows NT Server and Workstation provide complementary server and desktop networking services. NT Server offers simple installation, flexibility, and scalability. Integrated Internet, Intranet, and communications services enable the utilization of the Web for remote management and troubleshooting. The NT Server and Workstation packages support the same icon and desktop interface applications, and 32-bit API programming model as seen with Windows 95 and 98. As a result, Windows NT allows migration for existing users of Windows products.

Support for 32-bit addressing allows Windows NT to support a maximum of 4 gigabytes (Gb) of memory. With this, Windows NT provides improved performance by allowing more data and programs to reside in main memory. In addition, support for 32-bit arithmetic operations also improves the performance of some computational tasks.

The Windows NT support of Symmetric Multiprocessing, or SMP, allows multiple processors to share a given workload. To accomplish this, SMP provides a modular mechanism for central processing unit (CPU) capacity that increases through the addition of processors. Application multithreading breaks an application process into multiple tasks or threads with each thread capable of executing on a different CPU. As a result, application tasks can run concurrently. Because the operating system also uses threads internally, it also allows multiple instances of operating system functions to run concurrently.

Rather than use cooperative multitasking, Windows NT relies on preemptive multitasking. Cooperative multitasking uses a scheduler to control which process executes and the order of execution. Because the scheduler cannot add another process until the running process gives up the CPU, a higher priority process may wait for a lower priority process. Preemptive multitasking allows a high-priority process to preempt a lower priority process at any time. As a result, critical applications receive the required processing time, and the system remains responsive to on-line users or users running high-priority applications.

Windows NT also provides features for improving system availability. To reduce the chances for software-related system crashes, the operating system establishes memory and resource protection. With this, a faulty process cannot corrupt data belonging to other processes or the operating system. Windows NT also supports the automatic dumping of server memory or workstation registers to disk and rebooting if a fatal system error occurs.

Storage device protection occurs through RAID level 1 disk mirroring and the writing of data in stripes across multiple disk drives. In combination with an Uninterruptible Power Supply (UPS), Windows NT provides power-fail protection as a method of preserving data integrity. When the UPS generates signals that indicate a power failure, the Windows NT UPS services inform users that the system has begun to run off the UPS and allows them to shut down applications.

Windows NT offers a centralized management environment that allows the management and backup of distributed systems from a single location. With NT server administration tools, a network manager may work through any Windows client to manage any NT server on the network. The combination of the Windows NT Performance Monitor, Service Control Manager, User Manager, and Disk Administrator simplifies management tasks. While the Service Control Manager starts and stops services, the User Manager allows the addition, deletion, or modification of user information. The Disk Administrator provides a method for creating and deleting disk partitions.

Windows NT uses a centralized configuration Registry that contains operating system, application, and hardware configuration data. The Registry eliminates the need for configuration files and improves administrative efficiency by maintaining and updating from one location. To ensure the safety of critical information contained within the Registry, Windows NT either updates the entire Registry or does not update at all.

In addition to offering a centralized management environment, Windows NT also simplifies management through the support of the Dynamic Host Configuration Program, or DHCP. The DHCP uses a managed pool of IP addresses and dynamically assigns the addresses to network nodes. Microsoft Windows NT also includes support for Domain Name Server, or DNS, addresses within its Uniform Naming Convention, or UNC.

The support for Internet-related options also includes the inclusion of DNS server software, multiprotocol routing, a DHCP relay agent, and the Point-to-Point Tunneling Protocol, or PPTP. With the implementation of the Point-to-Point Tunneling Protocol, Windows NT enables a dial-up Remote Access Server client to access a Windows NT Remote Access Server securely through the Internet. Given this capability, a user could access a corporate LAN by dialing his local Internet Service Provider.

Microsoft Windows 2000

Microsoft Windows 2000 provides a multipurpose operating system that includes support for client-server and Peer-to-Peer networks. The Windows 2000 platform consists of the Windows 2000 Professional, Windows 2000 Server, Windows 2000 Advanced Server, and Windows 2000 Datacenter Server versions. As a whole, Windows 2000 offers the automatic installation and upgrading of software applications. In addition, the Microsoft operating system includes the capability to authenticate users before they gain access to resources or data contained within the individual workstation or the network. Windows 2000 also provides local and network security as well as auditing for files, folders, printers, and other resources.

With the use of Active Directory Services, Windows 2000 Server, Windows 2000 Advanced Server, and Windows 2000

Datacenter store information about network resources such as network accounts, applications, print resources, and security information. In addition, the Server packages permit users to gain access to resources throughout the network and to locate users, computers, and other resources. Along with Windows 2000 Professional, the Windows 2000 Server packages provide integrated support for most popular network protocols including TCP/IP and provide connectivity for Novell NetWare, UNIX, and AppleTalk. While Windows 2000 Professional supports one inbound dial-up networking session, the Server products support 256 simultaneous inbound dial-up sessions. The software also integrates user's desktops with the Internet and ensures secure Internet access.

Windows 2000 Professional offers a high-performance secure network client computer and desktop operating system. The package incorporates features from Windows 98 and extends the manageability, reliability, security, and performance of Windows NT. Windows 2000 Server operates as a file, print, applications, and Web-server software platform and works well for small- to medium-sized organizations. In contrast to Windows 2000 Server, Windows 2000 Advanced Server provides higher level network operating system services. Windows 2000 Datacenter supports large data warehouse, analysis, and simulations through the use of powerful server functions.

IBM LAN Server Aand OS/2 Warp Server

The LAN Server provides resource sharing plus facilities for defining, controlling, and managing access to LAN resources. The IBM product supports the NetBIOS, SDLC, token ring, Ethernet, and NDIS protocols and can allow access to a maximum of 1016 simultaneous users through a single server. The LAN Server also includes services that administer multiple servers as a single logical unit, DES compliant passwords, access control profiles for resources, audit trail capability for user-selected resources, and remote administration of printer queues. In addition, the LAN Server offers 32-bit Application Programming Interfaces for enterprise networks and extensions for Global Directory, security, and remote procedure call.

Introduced during 1996, OS/2 WarpServer combines the OS/2 WarpServer environment with the LAN Server to produce a fast file and printer server operating system. The software utilizes an object-oriented user interface, or OOUI, that allows easy installation, configuration, and system management. During installation, the software autodetects hardware and configures network adapters. In addition, WarpServer provides disk mirroring, remote administration, a backup server, and software metering.

UNIX

Originally introduced by AT&T, UNIX has become a multivendor operating system. UNIX provides a function-rich operating system that offers scalability from the desktop to the supercomputers. Many of the standards introduced through UNIX — such as mail, FTP, TCP/IP, and the domain name service — have become Internet standards. Because no broadly supported binary standard exists for UNIX, versions of UNIX may vary. As a result, the porting of UNIX applications from platform to platform often requires the recompiling of the application. Along with SCO, major vendors for UNIX include Sun, Hewlett-Packard, and IBM.

SCO Open Server

Introduced by the Santa Cruz Operation, the SCO Open Server product family comprises a complete line of advanced server operating systems for multiuser, networked, and enterprise solutions. The family provides a UNIX operating system for Intel processor-based systems to combine the power of

Product Group	Description
Base operating system	Internet server, multiuser enterprise server system, host server system, single-user desktop system for data-intensive client applications
Layered products	Windows services, Distributed Computer Environment products, reliability, availability, and serviceability add-ons, Internet services
Expert services	On-site engineering and consulting services, technical support, education services, on-line information services

TABLE 1.1 — SCO OPEN SERVER PRODUCT GROUPS

UNIX with a common and inexpensive platform. As shown in table 1.1, SCO Open Server includes the base operating systems, layered products, and expert services.

SCO Open Server Operating System Products

SCO Open Server encompasses four operating system products. The SCO Open Server Enterprise System, SCO Open Server Host System, and the SCO Internet FastStart products operate as server systems. In contrast, the SCO Open Server Desktop System works as a client system. The SCO Open Server Enterprise System performs transaction-based database management and business applications along with communications, file and print, and mail and messaging operations. Given the utilization of integrated graphics, multiprotocol networking, Internet services, mail and messaging applications, and remote systems, departments, retail stores and banks, and small-to-medium sized businesses use the SCO Open Server Enterprise System.

The SCO Open Server Host System runs transaction-based data management systems and business applications in a stand-alone environment or as part of a host network. Because the system targets environments that rely on a host computer, it provides a migration path to SCO Open Server

Enterprise System. As with the Enterprise System, the Host System includes integrated graphics, PC connectivity options, and mail and messaging services.

The SCO Internet FastStart System installs as an Internet server and includes a single-user version of the SCO Open Server Enterprise System, Netscape Navigator, and Netscape Communications Server. In addition to including the TCP/IP, IPX/SPX, NFS, NIS, DNS, PPP, SMTP, POP, and IMAP protocols, the Internet FastStart System features an HTML-based tool for installation and configuration. As already mentioned, the SCO Open Server Desktop System runs client-side applications. The Desktop System operates as a single-user desktop, offers RISC workstation capabilities, can multitask, and enables file and resource sharing.

SCO Open Sever Interoperability

Even though the SCO UNIX Operating System serves as the foundation of the SCO Open Server product line, servers equipped with SCO Open Server can provide a complete range of Microsoft Windows, XENIX, X Window Systems, and UNIX system applications to networks of personal computers and UNIX system workstations. The software can configure as a multiuser server for network users and locally attached character-based terminals, a multiuser database server, and a communications gateway server for local and remote users needing access to legacy systems. The Motif graphical user interface gives SCO the look and feel of Microsoft Windows.

Because SCO Open Server supports TCP/IP and Novell connectivity, it allows the interoperation between minicomputers, LANs, and other UNIX systems. Networking services for the Open Server Network System include full TCP/IP connectivity for remote logins, file transfers, remote printing, and client-server computing over asynchronous and LAN connections. In addition, SCO offers Network File System client and server support that provides distributed file systems services over TCP/IP. Novell IPX/SPX protocol support offers support for remote logins and client-server computing in a Novell environment.

UUCP and CU support provides asynchronous e-mail, file transfer, remote job-entry, and remote-login features to other UNIX systems. Support of the Post Office Protocol, or POP, allows POP-enabled clients to communicate with POP servers for mail management. The support of the Serial Line Internet Protocol, or SLIP, and PPP drivers allows SCO Open Server to support synchronous gateways or bridges.

2

Network Management

Introduction

Although the physical location of the personal computers on a local-area network seldom changes, networks have dynamic operations because the logical makeup of any network may change depending on the application. For example, the number of data and application files in use or in storage, the amount of available disk storage space, the number of users logged in to the network, and the volume of traffic passing through the network cabling can exhibit continual change. In addition, network users can take advantage of a distributed-processing environment where a centrally located server performs part of the processing and the other part of the processing occurs at the individual workstation.

In the most basic definition, network management involves maintaining the efficient use of network hardware, software, and media while providing consistent service for network users. Managing a network can range from the simple to the complex and may include reattaching a misplaced network cable, backing up network files, defragmenting disk directories, or searching all day for a solution to an operating system problem.

Network Management Requirements

At a basic level, network management requirements generally fall into five functional areas: configuration management, fault management, security management, performance management, and accounting management. Configuration management applications deal with installing, initializing, booting, modifying, and tracking the configuration parameters or options of network hardware and software. Fault management tools show a network manager the number, types, times, and locations of network errors by providing an audit trail of the errors such as dropped packets, retransmissions, or lost tokens.

Security management tools allow the network manager to restrict access to resources ranging from the applications and files to the entire network. Generally, security management tools offer password-protection methods that give users different levels of access for different resources. Security management also limits the capability to reconfigure network devices to individuals working at the systems administrator level.

Performance management tools produce real-time and historical statistical information about the operation of a network. As an example, a performance monitor may provide information about the number of packets transmitted at a particular time, the number of users logged into a specific server, or the use of internetwork lines. To accomplish these tasks, performance management tools poll individual network devices for component-specific information such as the data throughput for each serial port, the number of users logged in at a file server, the type of applications used, and the number of active files. Accounting management applications provide information that assists with the allocation of the costs of various network resources per user. This type of management process involves the compiling of information about the starting and stopping of sessions, user logins and resource use, and audit trail data.

Quality of Service

Quality of Service, or QoS, establishes parameters, or metrics, that guarantee network operation, availability, throughput, and response time. Some QoS strategies develop metrics specifically for a particular technology or protocol such as Ethernet networks, Storage Area Networks (SAN), or Asynchronous Transfer Mode (ATM) networks. In contrast, other QoS schemes remain specific to the IP protocol. Table 2.1 lists different QoS strategies.

Strategy	Description
ATM and Frame Relay Traffic Shaping	Bandwidth reservation mechanisms built into the ATM and Frame Relay standards.
IEEE 802.1p and 802.1q	IEEE specifications that allow Level 2 switches to provide traffic prioritization over Ethernet and Token Ring LANs.
Differentiated Services	Definition of ways to assign specific service levels and priorities to IP traffic using the IP TOS field.
Multiprotocol Label Switching	Method of encapsulating and tagging IP traffic to improve efficiency and control of routed networks.
Resource Reservation Protocol (RSVP)	Definition of how routers and other network devices should reserve bandwidth across the network on a hop-by-hop basis.

TABLE 2.1 — QUALITY OF SERVICE STRATEGIES

Network Backup

Performing a system backup is not the most exciting task and can seem like a waste of time. After all, new systems are much less likely to fail. However, when the new system crashes because of a component failure or a virus and the restoration of an important file is impossible, not having a backup will add more excitement to your life than anyone needs. Companies that lose critical data suffer loss because of the time needed to restore the data. In some cases, it may take hundreds of hours to recreate lost records and files.

All of us understand that the loss of important data will cause our computer systems and our services to grind to a halt. The *Disaster Recovery Journal* ran a survey that reported annual lost data costs of between $350,000 and $11 million per organization. Studies show that 80 percent of businesses that experience a serious data loss never reopen. A business insurance policy will pay to replace computer hardware, but it cannot replace what's inside the system.

On average, a business absorbs a cost of 50 cents when reentering a lost record or file. Let's perform a quick calculation. We can store roughly 1 million standard files on one gigabyte (GB) of hard drive space. Suddenly, our calculation shows a reentry cost of $500,000 per gigabyte. On average, a tape drive data protection system costs 0.1 percent of the value of the data.

Another *Disaster Recovery Journal* survey on system downtime reported that an average of nine occurrences per year per organization resulted in an average income loss of nearly $3 million per year. The reentry cost for 1,500 customer records at 50 cents each equals the cost of a tape drive that has the capacity to protect 4,000,000 customer records or standard files. Performing consistent system backup routines ensures that we can restore original data, user applications, and system configurations in a timely and efficient manner.

A Backup Plan

When establishing a plan for backing up a network, the plan should address the type and location of the system, node

Network Type	Network Speed	Megabits per Second	Gigabits per Hour
Ethernet	10	1.25	4.39
Fast Ethernet	100	12.50	43.95
FDDI	100	12.50	43.95
ATM	155	19.38	68.12

TABLE 2.2 — NETWORK TECHNOLOGIES AND BACKUP SPEEDS

names, IP addresses, and the network topology. The network topology stands as the major limiting factor in any client-server environment. As table 2. 2 shows, the difference in network technologies can affect the amount of data included in the backup.

The type of storage device associated with the backup also affects performance. Products based on Redundant Arrays of Inexpensive Disks, or RAID, offer the highest performance for on-line disk storage devices. With RAID, multiple disk drives can accept the transfer of data. Table 2.3 lists the RAID configurations.

Configuration Type	Description
RAID Level 1	Performs data mirroring that stores duplicate copies of data on separate disk drives. Each drive requires a mirror drive.
RAID Level 2	Performs disk striping. Multiple disks are set aside for error correction.
RAID Level 3	Stores data in parallel across multiple disks and supports applications that require high data transfer rates. All the drives in the array are used to fulfill a request every time a file is retrieved.
RAID Level 4	Stores and retrieves data using independent writes and reads to several drives. Multiple users can retrieve files at the same time but multiple writes are not possible.
RAID Level 5	Interleaves user data and parity data and distributes the data across several disks. The configuration works well for applications that require a high number of input/output operations per second. RAID Level 5 disks have a performance penalty during write operations.
RAID Level 6	Improves reliability by implementing drive mirroring at the block level so that data mirrors on two drives rather than only one drive. This configuration requires greater disk capacity.
RAID Level 7	Allows each individual drive to access data as fast as it possibly can. RAID Level 7 can support up to 12 host interfaces, can protect against a maximum of 4 disk failures, and can support multiple standby drives.

TABLE 2.3 — RAID CONFIGURATIONS

Magnetic tape provides a low-cost choice for long-term data storage. While the quarter-inch cartridge, or QIC, format accommodates a capacity from several hundred megabytes without compression to several gigabytes with compression, newer standards can store from 2 to 5 gigabytes without data compression. Although traditional tape drives have slow speeds, Digital Linear Tape drives provide increased storage capacity and performance. Tape arrays employ RAID technologies and multiply the speed and capacity while introducing fault tolerance.Many systems administrators have moved to optical disk technologies as a network backup solution. The optical options include Write Once-Read Many storage devices, magneto-optical drives, CD-ROM drives, CD-Erasable drives, and recordable compact disks. With the introduction and popularity of DVD drives for computers, interest in the use of optical backup devices has increased.

In addition to storage device types, the backup plan should consider the amount of total data contained within the system and set for backup. Any type of backup plan must also consider the plans that an organization has for network storage growth. The plan must also address the amount of time allocated for the backup process and whether backing up the network becomes a task for only the system administrator or for individual users.

Backup software controls all elements of an effective backup including scheduling, managing the placement of files on media, and providing fast and easy restore operations for the end-user or systems administrator. The software should provide a fast-start capability that allows full network backups to begin immediately. Moreover, the software should support multiple file systems including NetWare, Windows, UNIX, the Apple File Protocol, OS2 High Performance File System, the Network File System, and OSI File Transfer, Access, and Management.

Most backup software packages also provide a fast search capability at the file level, automatic file and directory storage, the capability to delete data not accessed for a specified period of time, and the capability to select either a full, incremental, or differential backup. Typically implemented on a nightly basis, incremental backups copy only files that have changed since the last backup. Although the practice of incremental backups speeds the operation, it also requires more detailed management practices because each backup media may contain different files. Differential backups require media with a full set of files. However, each media set used for the differential backup contains all files that have changed since the last full backup rather than the last incremental backup.

Most client-server backup software packages feature a client agent that initializes at a scheduled time and a server agent that collects the data and places the data on tape while starting all client operations. The scheduling of backups should consider the critical nature of applications and network availability. In addition, the software should provide overwrite protection, log file analysis, and media labeling. Some backup software remotely mounts the client file system; other software packages allow multiple clients to process backups concurrently.

Any backup operation requires substantial CPU power for the processing of data into and out of a computer. Depending on the backup software, varying transfer rates may occur. The file system and file sizes will affect backup performance. Small file sizes range from 0 to 64 kilobytes (Kb) will process at a slower rate because of the smaller input/output tasks at the disk and the constant need for repositioning of the read/write head.

Implementing a Disaster and Restoration Plan

Local-area networks operate as data-intensive environments that require special precautions to safeguard integrity and reliability. Along with scheduled backups, the precautions also include the implementation of disaster and restoration planning procedures and the purchase of equipment that provides redundant protection against failure. Usually based on the Simple Network Management Protocol and Remote Network Monitoring, the use of a solid network management platform that can monitor network traffic and component problems can prevent problems from occurring. Redundant equipment activates automatically when various network devices fail, prevents data loss, and maintains network reliability.

Reliability

Reliability addresses the ability of a network to continue operating despite the failure of a critical element. In most definitions, reliability involves redundancy and could result in the doubling of component and media costs. When comparing network topologies, the star topology offers better reliability than the ring topology. The loss of a link in the star topology results in a communication failure between the hub and the affected node while other nodes continue to operate. However, the failure of the hub will cause the network to fail. In the ring network, each node remains actively involved in the transmissions of other nodes. Because each node depends on the reliability of other nodes, the loss of a link causes the network to fail.

Availability

Availability measures the performance of a LAN in terms of user accessibility. A network that has high availability provides immediate services to users while a network with low availability forces users to wait for access. Most administrators implement a management and monitoring system that can detect high network utilization and high error rates as a method for achieving good availability. In addition, the installation of intelligent hubs can provide good availability for a network. An intelligent hub provides circuit redundancy, failed circuit bypass, and centralized monitoring while concentrating the topology in the backplane of the hub. As a result, a systems administrator can implement bypass procedures without shutting down the network.

Recovery and Reconfiguration

Recovery involves the restoration of LAN operation back to a stable condition after an error occurs while reconfiguration restores operation after a failure. When considering the purchase of recovery software, functions should include the capability to restore service or power, to control packet loss, and to prevent transmission collisions, bit errors, noise, and service faults. The recovery or reconfiguration of the network after faults or failures requires the capability to detect and isolate errors, determine the effect of the errors, and indicate actions that can correct problems.

Reconfiguration should restore service after the loss of a link or network interface unit and usually involves the bypassing of major network components. The configuration assessment portion of a reconfiguration system unlocks historical information kept in logs and uses the information about current connectivity, component placement, and path configuration to restore normal operation. The system provides a method for the system administrator to assess how a particular failure affects the system.

Solutions may involve the reconfiguration of most operational processes as a requirement for avoiding the error. The solution determination component of the reconfiguration software examines the configuration and affected components, determines the best method for moving resources, indicates portions of the system for possible elimination, and identifies network components for service. To accomplish these tasks, the solution determination component defines the priority of critical network functions and includes all available network resources for the solution.

In some instances, the reconfiguration system can apply this information toward the identification and implementation of an alternate configuration. The use of an alternate configuration may require the rerouting of transmissions, the moving and restarting processes from failed devices, and the reinitializing of software. If a failure does not involve a critical network function, the reconfiguration system may notify affected users and take no other action.

Distributing Resources at the Server

The addition of multiple special-purpose servers provides users with functionality, connectivity, and processing power that a single file server and network operating system cannot offer. As an example, a single multiprocessor server combined with a network operating system can provide enough throughput to support five to ten times more users and applications than those operating systems supported by a microcomputer operating as a server. The distribution of resources minimizes the disruption to productivity that results from a failure. Compared to a LAN with centralized resources, a LAN that uses specialized devices as servers permits the integration of diagnostic and maintenance capabilities and offsets the range of failures that can occur.

Fault Tolerance at the Server

Because a failure at the server can result in lost or destroyed data, network administrators usually commit the most resources at the server. Depending on the desired level of fault tolerance, an administrator can implement unmirrored, mirrored, or duplex configurations at the server. An unmirrored configuration uses one disk drive and one disk channel including the controller, a power supply, and interface cabling. A failure in either the drive or the disk channel can cause the temporary or permanent loss of stored data.

A mirrored server uses two similarly sized hard disks and one disk channel. The two disks mirror together over the same channel and controller. As data writes to one disk, it automatically copies to the other disk. If one disk fails, the other disk becomes active, protects the data, and ensures that all users have access to the data. However, the failure of the disk channel or controller causes both disks to become inoperative. Because the disks share the same channel and controller, the writing of data to the disks occurs sequentially and can slow network performance.

Shown in figure 2.1, disk duplexing involves the installation of multiple disk drives with separate disk channels for each set of drives. If a malfunction occurs anywhere along the disk channel, the remaining channel and drives continue to operate normally. Because each disk drive uses a separate disk channel, write operations occur simultaneously. Both drives receive the same read requests; the drive that has the closest proximity to the information responds to the request while the second drive discards the request. Duplexed drives can share multiple read requests for concurrent accessing of data.

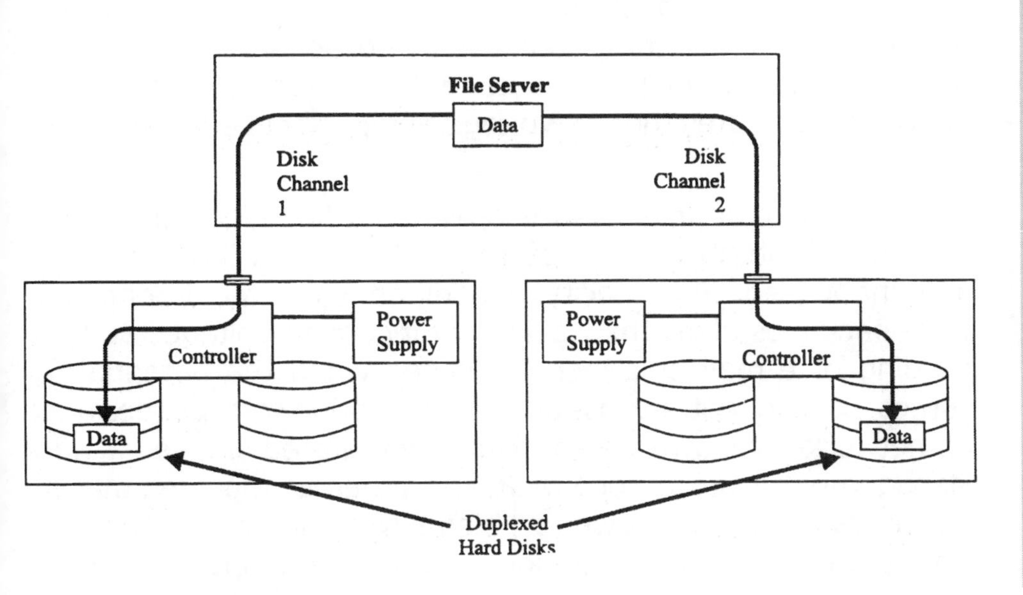

Figure 2.1. Disk-duplexing configuration

Uninterruptible Power Supplies

Uninterruptible power supplies, or UPS, protect computer systems by continuing to supply power to the system after a power outage. In addition, a UPS uses a battery backup and switches the computer to the battery if the ac voltage surges high or sags low. All these precautions prevent data and file corruption and possible damage to computer and networking hardware and peripherals. The actions provided through the UPS give users enough time to shut applications down properly and to save files.

Uninterruptible power supplies may appear as off-line, on-line, and interactive technologies. The off-line, or standby, UPS units provide benefits if a power outage occurs and switch the computer system to the battery backup through an inverter. On-line UPS technologies use an inverter that continuously converts the alternating current (ac) to direct current (dc) for the battery and back to ac for the computer. Because the on-line UPS operates continuously, it protects against surges, sags, and outages. Line-interactive UPS technologies feature a continuously operating inverter that charges an internal battery but that does not constantly convert the direct current to the ac required for the computer. The dc-to-ac conversion occurs only in the event of a power outage. Line-interactive UPS hardware also regulates the line voltage and protects against surges and sags.

UPS monitoring software provides information about the amount of battery runtime stored within the UPS unit and the amount of used UPS capacity. In addition, some monitoring software packages allow the tracking of incoming voltages and the sending of e-mail or pager messages when the potential for surges, sags, or battery loss occurs. Monitoring software designed for networked hardware also includes integrated Simple Network Management Protocol, or SNMP, and Ethernet modules that allow users to monitor power problems from any node. The software stores monitoring data into a database for later analysis.

Network Management Tools

Network management tools include network analyzers, application-specific tools such as a performance monitor, and proprietary software such as IBM's mainframe-based NetView and give network managers information needed for the maintenance of network operation. In brief, the tools and equipment associated with network management may find cable breaks, pinpoint the cause of a network slowdown, preserve predetermined performance levels, manage network resources optimally, maintain security, track network use, and provide a basis for charging customers for network time. For example, information about workstations that generate the heaviest traffic can allow a network administrator to prevent possible bottlenecks through the addition of internetwork devices such as bridges or routers.

Management Through Software

Most network-attached hardware devices support SNMP, RMON, Java, or ActiveX software agents that provide current status information to a network manager or maintain traffic statistics for subsequent analysis. The Java and ActiveX software agents allow the development of interactive, browser-based applications that can run on existing systems.

Simple Network Management Protocol

As networks have grown larger and become increasingly heterogeneous, the need for industry-standard network management protocols and products that operate across a wide range of vendor offerings has also grown. Developed by the Internet Activities Board during 1988, the Simple Network Management Protocol relies on the User Datagram Protocol/Internet Protocol, or UDP/IP, as the underlying mechanism for transferring data between different types of systems and networks. SNMP defines the communication between a network management station and a device or process set for management. Originally developed as a lightweight, practical protocol for managing bridges and routers in networks using TCP/IP, SNMP has become the most widely implemented nonproprietary management solution in the world because of low-cost implementation and the capability to support vendor specific extensions.

The three-layer architecture of SNMP consists of network management stations, agents, and a common set of protocols that binds them together and operates with a management information base, or MIB, and a structure of management information, or SMI. The MIB and SMI network management concepts allow the defining of each network element and the monitoring and controlling of the elements by management stations.

SNMP gathers management data about device status through polling across the network. A streamlined set of commands based on the GET and SET operations establish the polling processes. GetResponse and GetBulk allow a management station to inspect MIB variables, and SetRequest allows a management station to alter the variables. SNMP agents may send unsolicited messages using the Trap-PDU command to guide and focus the polling.

Remote Monitoring Standard

A remote monitoring standard called RMON defines a structure for storing data gathered from probes that sit on each segment of a LAN. With this, any device that includes an appropriate agent module can have all the features of an embedded LAN performance analyzer with the additional benefits of integrated hub management and control. The probes view every packet passing through the LAN and produce summary information on various types of packets such as undersized packets and packet collisions. In addition, the probes store information for further analysis by capturing packets according to predefined criteria set by the network manager.

Group Name	Group Description
Filters Group	Provides a buffer for incoming packets and user-defined filters.
Statistics Group	Maintains low-level usage and error statistics such as the number of packets sent, broadcasts, and collisions.
History Group	Provides user-defined trend analysis based on information obtained from the statistics group.
Host Table Group	For each host, contains counters for broadcasts, multicasts, errored packets, bytes sent and received, and packets sent and received.
Host Top N Group	Contains sorted host statistics such as a complete table of activity for the three busiest nodes communicating with each host.
Alarms Group	Allows a sampling interval and alarm threshold to be set for any counter or integer recorded by the RMON agent.
Packet Capture Group	Allows buffers to be specified for packet capture, buffer sizing, and the conditions for starting and stopping packet capture.
Events Group	Logs events such as packet matches or values that rise or fall to user-defined thresholds.
Traffic Matrix Group	Arranges usage and error information in matrix form to permit the retrieval and comparison of information for any pair of network addresses.

TABLE 2.4 — RMON MIB OBJECT GROUPS

The RMON management information base includes nine object groups and associated variables. Table 2.4 provides a listing and description of the object groups.

ActiveX

ActiveX allows the development of applications with Windows-based tools such as C++ and Visual Basic. Because ActiveX has a basis in Microsoft's Object Linking and Embedding, or OLE, and Component Object Model, or COM, technologies, OLE Server objects such as Lotus Notes, Microsoft Office, Visio, and Microsoft Internet Explorer can host ActiveX applications. The ActiveX Control Pad utilizes a point-and-click interface and offers an authoring tool that allows the addition of ActiveX-based content to existing Web pages. ActiveX relies on system-level services provided by Windows 95/98, Windows 2000, and Windows NT and resembles Microsoft's Win32 API.

Java

Introduced during 1996, Java provides an interpreted, machine-independent programming language. As with ActiveX, Java gives standard HTML-based Web pages intelligence. Java programmers create compact and highly functional applications called applets that have a basis in a macro/scripting language. With the use of the scripts, a user can create Web pages that have automatic search functions or compile data. Because Java features platform independence and because browsers include Java support, Java applets can run on any type of equipment.

Hardware Management Tools

Hardware tools for network testing include dedicated equipment such as protocol analyzers, breakout boxes, and servers established for the purposes of network management, testing, and tuning. The implementation and degree of functionality available in test equipment varies from product to product. Developing end-to-end testing capabilities that simulate new applications on an Intranet may require a basic toolset that covers protocols, configuration testing, and network mapping.

Baselining

Baselining involves the process of capturing snapshots of a normally operating network and determining the normal values of key network statistics. After establishing a normal behavior baseline, a network administrator can compare the baseline with ongoing network behavior. As a result, spotting abnormal network behavior and preventing network problems becomes much easier.

Baselining also serves as the first step toward tuning, or optimizing, network performance. Tuning refers to improving performance without purchasing more equipment or ordering more bandwidth when bottlenecks occur. Network managers who establish baselines and study traffic patterns can achieve performance improvements by relocating servers to the same segments as the heaviest users or by splitting and rerouting protocol traffic. Because each network has a unique mix of components and user traffic patterns, each network also has a different set of measurements that indicate normal operation and require the study of patterns rather than the simple comparison of measurements.

Baselining also facilitates network capacity planning. By reviewing weekly and monthly summaries, network managers can spot gradual increases in network utilization and anticipate bottlenecks. All this attention allows for the tuning and upgrading of the network in anticipation of higher capacity requirements. A history of measurements allows managers to predict when segments will reach capacity and determine whether further segmentation will alleviate the situation. By comparing overall growth with the growth of traffic by specific protocols, network administrators may determine whether splitting protocol traffic will improve performance.

Most network administrators rely heavily on commercially available traffic monitors and protocol analyzers to collect data for the baselining processes. Although protocol analyzers provide a full seven-layer decoding of protocols, traffic monitors perform long-term historical analysis and the capability to simultaneously view multiple segments. Most administrators will place a traffic monitor on each critical LAN segment.

Using Protocol Analyzers for Network Management

During typical network operation, communications software formats data for transmission according to a particular protocol by adding a header and trailer and forming an envelope for the message. Devices on the network using the same protocol can read the envelope, route it over the appropriate link, deliver the data to the appropriate addressee, and provide an acknowledgment. A protocol violation occurs when the network or device does not follow established procedures. From there, a technician must locate the problem before restoring communications across the network.

Protocols exist as either bit-oriented or byte-oriented. Byte-oriented protocols include bi-sync and poll-select. Bit-oriented protocols include IBM SNA/SLDC, Ethernet, Token Ring, and TCP/IP. Of the two types, bit-oriented protocols offer the most difficulty for analysis and require the use of specialized test equipment.

A protocol analyzer views and verifies the protocol transfer process and decodes messages. Protocol analyzers exist for all types of communications services including X.25, frame relay, ISDN, Token Ring, Ethernet, FDDI, and ATM networks. As shown in figure 2.2, a protocol analyzer connects directly to the network in the same way that a node or communications port connection exists.

In the passive monitoring application, the protocol analyzer displays the protocol activity and user data passing over asynchronous or synchronous transmission links. As a result, the analyzer establishes a window into the message exchange between network nodes. Most protocol analyzers have features such as data capture to random-access memory (RAM) or disk, automatic configuration, counters, timers, traps,

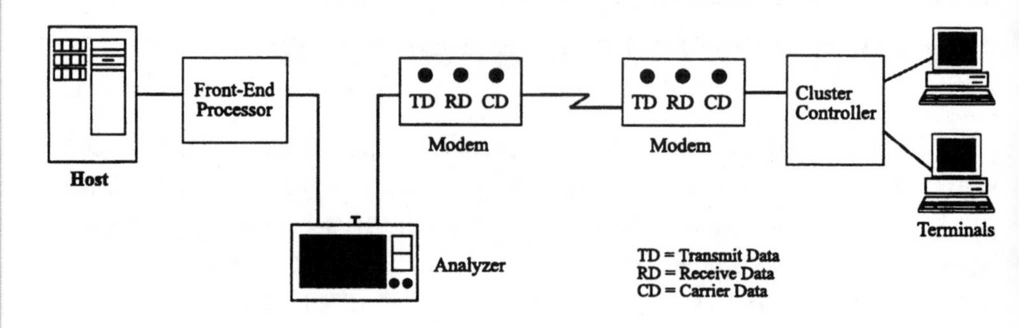

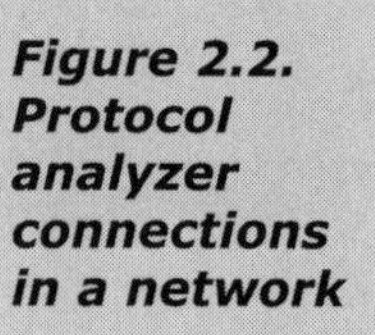

Figure 2.2. Protocol analyzer connections in a network

masks, and statistics that assist with reducing the time needed for diagnosis. Trapping provides a method for recording only essential data into the buffer of the protocol analyzer or onto a disk drive. As an example, a trap could capture the first incorrect frame received at the analyzer. Filtering provides the capability to include or exclude certain types of protocol data such as the destination, source, or bilateral addresses, protocol type, or error packets. Postfiltering captures a large number of packets to disk or RAM for the purpose of further analysis.

Even though Bit Error Rate Testing, or BERT, capabilities vary from analyzer to analyzer, the range includes full-duplex, half-duplex, and multidrop support. With the generation of packets, a protocol analyzer allows for the impact of additional traffic on the network to be tested. Using a group of configuration screens, a network administrator can set parameters such as:

- source address,
- destination address,
- maximum frame size,
- minimum frame size,
- the number of packets sent out with each burst.

Packet generation also allows the customization of the data field section contents to simulate real or potential applications.

Load generation creates varying traffic rates on the network. As the network is loaded from lowload to overload conditions, the protocol analyzer stresses network components such as hubs, repeaters, bridges, and transceivers for the purpose of identifying weak links. The timers contained within protocol analyzers measure the time interval between events. By establishing a trap in the transmit path and another in the receive path, the protocol analyzer can verify the handshake procedure time interval.

Mapping automatically documents the physical location of LAN nodes and eliminates the need for rearranging the network map after devices are added, deleted or moved. In addition, mapping software allows for the naming of nodes and includes icons for servers and workstations. The icon provides information about the type of adapter used at the node and the location of the node along a cable run.

Analyzer software usually includes a text editor that can run with captured data. With this resource, a network administrator can delete unimportant data, enter comments, print reports, and create files. The text search function allows the tracking of a problem source in data packets through the identification of a known text string.

Along with the more sophisticated troubleshooting and monitoring functions, many protocol analyzers include a mechanism for testing cable breaks or improperly terminated connections. Time Domain Reflectometry, or TDR, sends a signal down the cable, receives and echo, and interprets the information. The analyzer may report that the cable has no fault detected, no carrier sense, an open on coax, or a short to coax.

In addition to the monitoring and troubleshooting tests, some protocol analyzers incorporate sophisticated programmability and simulation options. Some analyzers require the use of a programming language, whereas others use a setup screen that allows for the definition of a sequence of tests. As soon as conditions occur that meet predetermined thresholds, the program automatically initiates a set of tests used for tracking and solving the problem. An automatic configuration capability allows a protocol analyzer to configure itself automatically to the protocols found at the line under test.

With the simulation application, programming of the protocol analyzer allows the device to operate as a gateway, communications controller, or front-end processor. The protocol analyzer replaces the suspect device on the network and al-

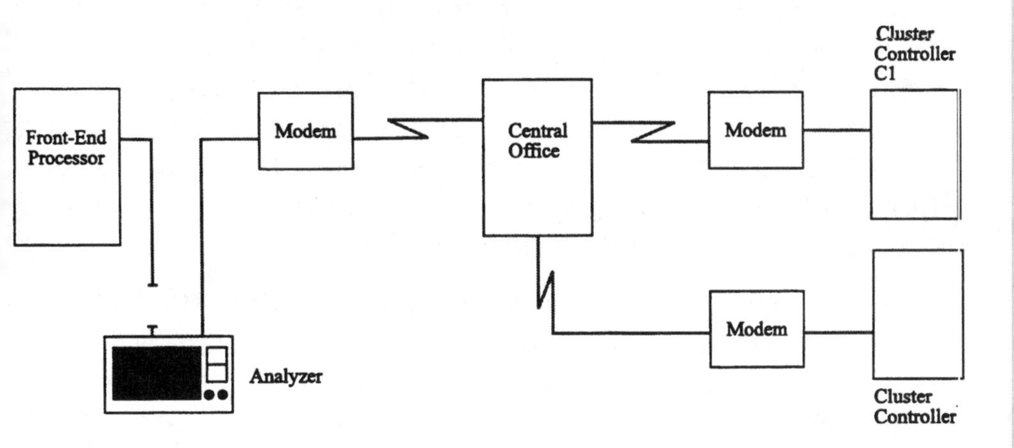

Figure 2.3. Protocol analyzer in simulation mode

lows for the isolation of problems. More sophisticated simulation routines verify the conformity of a network device to specific standards. Figure 2.3 shows the connection diagram for a protocol analyzer operating in simulation mode.

Using Network Analyzers for Network Management

The network analyzer shown in figure 2.4 provides more functionality than a protocol analyzer and offers fault and performance capabilities for maintaining, troubleshooting, and fine-tuning networks. To accomplish these tasks, network analyzers provide the capability not only to oversee the seven network model layers but also to provide automatic analysis and alarms. The analysis information can transfer to a spreadsheet or database. In addition, the network

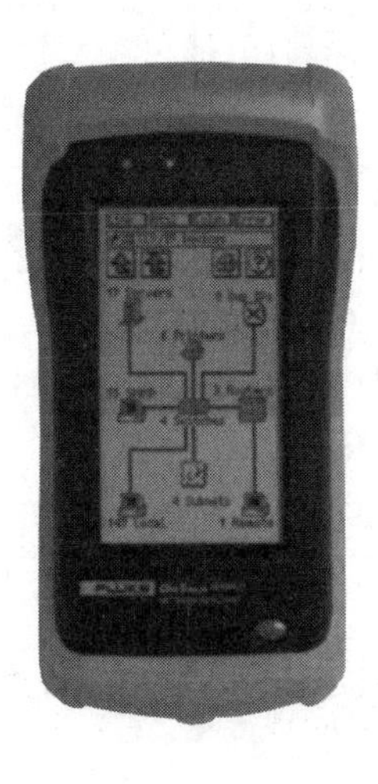

Figure 2.4. Network analyzer

analyzer can detect problems such as bottlenecks before an interruption to network services occurs. Network analyzers exist for a variety of LANs and WANs applications.

Breakout Boxes

A breakout box monitors the performance of a network without interrupting user traffic. By connecting a breakout box between two devices, a network administrator can check for the proper connection of interface leads. The breakout box allows the opening, closing, or cross connection of the leads for any pattern. Because a breakout box has the capability to cross-connect leads, it can connect devices that do not have identical interfaces. In addition, a breakout box can test cable continuity.

Using Cable Testers for Maintaining Network Efficiency

Most, if not all, cable testers offer ease of use. By using the tone generator often found in cable testers, a technician can plug the tester into the cable at the PC end, go to the wiring closet, and wave the tone probe over the cables. The ringing sound identifies the terminating end of the cable.

When conducting tests, one member of a two-person team works from the telecommunications closet, unplug a cable from the hub or patch panel, and attach the cable to the tester. The other member of the team connects the remote unit of the tester to the terminating end of the cable located at the user's work area. From there, the technicians can run an autotest, or a series of predetermined tests, and grade the system. Test results store within the testing unit. When moving to the next cable, the person at the terminating end moves to the next wall jack to repeat the process.

As we discovered in the opening sections of this chapter, the components and cabling that make up a telecommunications system must have uniform impedance values. Faults within the cable can change impedance. Consequently, a load with one impedance value will reflect or echo part of a signal being carried by a cable with a different impedance level and cause failures. Because of this scenario, both vendors and installers test to ensure that the impedance, resistance, and capacitance values found within the networking cabling comply with standard cable specifications.

Cable testers check for impedance mismatches during tests for cable faults. For example, a break in a wire creates an open circuit and infinitely high impedance at the break. When a high-frequency signal emitted from a cable tester encounters this high impedance, the signal reflects back to the tester. In contrast, a short circuit within a cable displays as a zero impedance value. With this, the impedance mismatch also reflects a high-frequency signal. However, the signal reflected by the short circuit will have an inverted polarity.

Most cable testing devices can provide an approximate measurement of the distance to the cable fault. To do this, the tester relies on a cable value called the nominal velocity of propagation, or NVP. The NVP measures the rate at which a current can flow through the cable and expresses the measure as a percentage of light speed. Then, the cable tester multiplies the speed of light by the NVP and by the total time taken for the pulse to reach the fault and reflect back to the tester. To show the one-way distance, the tester divides the measurement by two.

When checking the electrical length of a cable installation, a tester applies the same concept by using Time Domain Reflectometry to measure length. However, this test requires that one end of the cable not have any termination. During the test, the open end will register as infinite impedance and reflect a pulse back to the tester. Then, the tester factors the response time into the formula used to estimate the overall electrical length of the wire. Cable testers do not have the capability to check the first 20 feet of a cable because a pulse transmitted by the tester reflects back to the device before the signal transmission concludes. As a result, the tester cannot provide an accurate measurement.

Noise stands as one of the biggest problems for a cabling system. A cable tester transmits different frequencies to check the ability of the system to dampen the effects of noise. Due to attenuation, the transmitted signals vary from high strength at the transmission point to lowest strength at the destination point. As a result, the magnetic field of a signal transmitted from a device through one wire may overwhelm a signal arriving at the same device on the wire pair.

Purchasing a Cable Tester

As network technologies have progressed to the high-speed transmission of data, network managers have faced the challenge of knowing whether projected or installed cabling systems will handle the increased data rates. Because of this, test equipment vendors have added a wider range of tests and tools for taking and interpreting measurements. New cable testers provide LCD display screens and cable grading and troubleshooting capabilities and also work as fault locators, Ethernet monitors, cable toners, and voice sets.

Given the new functionality, selecting the correct cable tester for the application also exists as a challenge. Much of challenge arises from the different test and evaluation philosophies exhibited by test equipment vendors. While some vendors point to the need to certify that a cabling system can handle specific high-speed applications, others emphasize testing to ensure that cabling exceeds current standards. Within this second set, vendors may define specific levels at a specific frequency or recommend the taking of measurements over a range of frequencies.

In addition to capabilities, network managers and technicians must also consider the physical size, portability, and cost of cable test equipment. As an example, hand-held cable testers used for testing Cat 5 cabling range in weight from one to more than three pounds. Figure 2.5 shows two popular hand-held cable testers, and figure 2.6 shows two network cable scanners. With and without various options, purchase costs for full-function cable testers range from $2,500 to $5,500. Smaller hand-held units that offer continuity testing rather than total test functionality may sell for as little as $110.

Figure 2.6. Network cable scanner

Figure 2.5 (far left). Cable meters

Cable Tester Specifications and Test Sets

With the major points listed in table 2.5, Telecommunications Systems Bulletin (TSB) 67 authored by the EIA/TIA establishes standards for cable testing. Every cable tester must have the capability to run a suite of four tests: NEXT, wiremap, length, and attenuation. Minimum performance requirements depend on the cable type undergoing test. In addition, TSB 67 also defines two degrees of accuracy defined as Levels I and II.

TSB 67 defines two test configurations called Basic Link and Channel. Basic Link testing covers the permanent portion of the cabling from the wall outlet to the first point of termination in the telecommunications closet. As a result, the test occurs before the installation of any network hardware. Channel testing covers the entire cable run including the cable, patch cords, and all connections.

Function	Description
NEXT	Measured on all six pair combinations from both ends of the link.
Wiremap	Verifies that the pin-and-connector pairs on either end of the link match as specified by the configuration.
Length	Verifies that the distance of the run does not exceed TSB 67 limits.
Attenuation	Verifies that the maximum attenuation value as defined in the cable specification is not exceeded.

Table 2.5 — TIA/EIA TSB 67 Required Tester Functions

Despite the different philosophies, a cable tester has basic functions and may work with various cable types. When purchasing a cable tester, verify whether the cable will accurately test all or only a portion of the cable types discussed in this chapter. In addition, check for the availability of a fiber-optic probe attachment and automatic bidirectional NEXT measurement. The tester should have the capability to process and store tests quickly. Some products may run an entire suite of tests within 8 seconds; however, other testers require up to 45 seconds for the same task.

Because networks feature multiple links, the ability to store and manipulate test results has become critical. Cable testers may store only the last test performed to as many as 2,000 tests. In addition, some cable testers allow the downloading of results to a personal computer so that installers can provide customers with a printed summary of each test link.

Function	Explanation
Powersum Next	PS-Next measures the signal coupled from all adjacent pairs. New technologies such as gigabit Ethernet use all four pairs.
Propagation Delay	The time required for electrical signals to travel from one end of the cabling link to the other.
Delay Skew	The difference in time required for a signal to travel the length of the link over each of the wire pairs.
Attenuation-to-Crosstalk Ratio	ACR measures the strength of the signal in relation to the noise on the same pair. A high ACR yields a lower the error rate and a higher quality connection.
PS-ACR	The ratio of attenuation to PS-Next.
Impulse Noise	Noise induced by nearby equipment such as light fixtures.
Capacitance	Measures the signal distortion caused by the interaction between electrons on two nearby wires.
Impedance	Measures opposition to the flow of electrical current. Impedance is measured for each pair in the cable.
Return Loss	The power of the signal reflections measured at the cable relative to the power of the transmission.
Loop Resistance	Resistance measured in a loop through each pair in the cable.
Test to Frequencies as High as 155 MHz	accurately predicts if a cable system can support 155-Mbps ATM traffic.
Cable Grading	Quantifies link performance against minimum Cat 5 requirements.
Measures Across Multiple Frequencies	Identifies cable that minimally performs beyond required specifications.
Measure Ethernet Traffic	Locates unused ports and measures utilization, collisions, and lengths of frames.
Tone Generation	Traces a particular cable from node to telecommunications closet.
Topology Autotest	Conducts tests according to cabling or network type.
Auto-Troubleshooting Mode	Gives a detailed analysis when any test in the autotest series fails.
Failure Location	Uses crosstalk analysis to pinpoint the location of a failure.
Upgradeability	Upgrades through either software or EPROM.

TABLE 2.6 — CABLE TESTER FUNCTIONS

Regardless of the speed, the true measure of tester usefulness is accuracy. Vendors may express accuracy in terms of maximum and typical accuracy levels. If a tester conforms to TSB 67 requirements, the device must provide an indication of a marginal pass or fail. If a cable passes any of the tests by a margin that is lower than the accuracy of the tester, a "marginal" rating occurs. Although most cable testers offer TSB 67 Level II accuracy for Basic Link tests, few offer Level II accuracy for Basic Link and Channel tests.

Manufacturers ship cable that can carry data at frequencies far higher than the 100 MHz defined in the Commercial Building Telecommunications Cabling Standard 568A specification. Cables carrying high-frequency signals become more susceptible to noise caused by signal reflections. Given the functions shown in table 2.6, many cable testers check the actual bandwidth of the cable and give an indication of how cable will handle high-speed technologies.

Capacity Planning

Network administrators must give more attention to accurate capacity plans for network traffic, host utilization, and application performance. In addition, the technology environment of today requires the continuous updating of the plans. Capacity planning becomes increasingly important because of the need to justify financial and time investments. Proper capacity planning can help identify potential bottlenecks before the problems occur and prevent most performance-related problems. The lack of good capacity planning for a network or a technology system can lead to user dissatisfaction, a decrease in technician productivity, budgetary problems, and stability problems with the system.

Without proper planning, performance may suffer at peak loads. As a result, network clients may wait an excessive amount of time for a response to a request. Moreover, systems that cannot handle the expected peak throughput will cause a loss of technician time. As an example, technicians may spend a significant time analyzing network bottlenecks that occur because of poor design.

Planning of any nature must consider short- and long-term budgetary impact. As a result, capacity planning may also require some consultation with accounting officers. In some cases, replacing or improving hardware stands as the only options when system performance begins to lag or when business tasks take a system to its operational capacity. With proper capacity planning, an organization receives the justification needed for expenditure and can budget for upgrades or purchases.

By identifying potential problem areas and capacity limitations, managers can avoid and possibly predict stability problems. As it happens, stability problems usually occur at peak user loads. Proper capacity planning allows organizations to identify periods when stability problems may occur and provides the time to prepare for or prevent a problem.

Financial considerations drive capacity planning. To effectively perform capacity planning, an organization must allocate time, internal and external staff resources, and software and hardware tools. In terms of costs and benefits, the lack of capacity planning may result in critical delays, the installation of more network equipment or services than necessary, and increased costs when upgrading a system already in production.

Quick changes in technology trends make capacity planning more critical. With the trend toward centralized servers rather than distributed servers, it becomes essential for the one server to perform well. In addition, the network interface for that server also has a critical function. Capacity planning also considers the potential for technological change and advancement. As an example, multimedia information in the form of streaming video and audio traffic has found a place in production network systems and will strain existing network infrastructures.

In the end, capacity planning asks and answers three basic questions from both the network side and the customer side. Those questions are:

1. What will the changes do to my network utilization?
2. How will the response time for my client's customers improve with a new application?
3. How will the new application affect performance of existing applications?

Proactive Capacity Planning Through Teamwork

A business manager must create an environment where information flows to and from system administrators, technicians, and clients. Without this type of communication, capacity planning will not work. Before recommending a technology application to any organization, a manager should understand the functions and potential uses for the application. To gain this understanding, managers can conduct interviews with systems administrators, technicians, and clients to learn about important transactions. For example, if a specific routine in a particular database queries 50 times an hour, then that query must take much less than 60 seconds. It is impossible to do network capacity planning without understanding what transactions are important to end users.

Setting a Sample Period for Measurement

Diagnostic tools and statistical reports allow the monitoring of the performance of specific applications. When using these tools and reports, consider the sample period for the measurement. A utilization report averaged over 15 minutes may miss many peaks that could affect a customer's and the client's perception about response time.

When determining the projected network utilization, consider the application along with the client and customer needs. Many systems provide methods for calculating network utilization over a set period of time. As an example, a system may provide analytical modeling tools that allow a projection of network utilization. Capacity planning also requires the careful monitoring of network traffic. Without this step, the accuracy or inaccuracy of your planning will remain unknown.

3

The OSI Model:

Physical and Data Link Layers

Introduction

Illustrated in figure 3.1, the Open Systems Interconnection (OSI) model consists of seven layers of protocols that cover the communication of raw data and the networking of applications. Looking at the model, the layers range from the Physical Layer, which covers the transportation of data, to the Application Layer, which contains user applications. With this, the OSI model establishes a reference for layered networking architectures. The model offers "open" standards because the interconnection between networks occurs without specifying any type of hardware. Instead, the communications software adheres to given standards.

To better understand the functionality of the OSI model, we can place actual network functions alongside each layer. Briefly, the bottom four layers of the model cover the communication of raw data, and the top three layers cover the networking of applications.

OSI Model Definitions

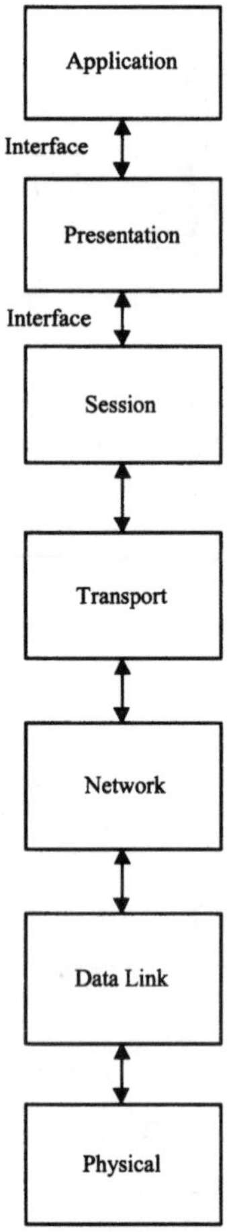

Application

Interface

Presentation

Interface

Session

Transport

Network

Data Link

Physical

Figure 3.1. Open system interconnection model

Physical Layer

The Physical Layer resides at the lowest layer of the OSI model and transfers bits of data across some type of physical link. Physical links may consist of twisted-pair cabling, coaxial cable, fiber-optic cabling, wireless radio signals, and satellite transmissions.

Data Link Layer

The Data Link Layer covers the transmission of frames and functions in tandem with the physical layer and the network layer. Data Link Layer operations may provide either connectionless transmission, which requires no acknowledgment, or connection-oriented transmission, which requires an acknowledgment of service.

Network Layer

The Network Layer sends packets of information and handles any problems associated with delivering a packet of information form one node of the network to another node. A process found within the Network Layer communicates with processes found at the other ends of all communication links connected to the transmitting node. During the interaction between the Network and Transport Layers, the Network Layer establishes a unified addressing scheme. Each node on the network has a unique address that becomes part of a total, consistent addressing scheme for the network.

The Physical Layer

The Physical Layer resides at the lowest layer of the OSI model and transfers bits of data across some type of physical link. The medium that makes up the physical link may consist of twisted-pair cabling, coaxial cable, fiber-optic ca-

OSI Model Definitions (continued)

Transport Layer

The Transport Layer sees the entire network and uses the Physical, Data Link, and Network Layers to establish end-to-end communications for the higher levels of the model. The Transport Layer has the primary task of moving messages from one end of a network to the other end. In addition, the Transport Layer also provides a set of network services that include addressing, connection management, data flow control, and buffering.

Session Layer

The Session Layer sets a number of parameters that allow actual communication between the nodes to occur. Before the transfer of information takes place, the Session Layer ensures that communication can occur and that the nodes do not attempt to communicate simultaneously. Then, the Session Layer manages the communication by breaking it into usable parts. When the communication ends, the Session Layer provides an orderly method for the break to occur.

Presentation Layer

As data flow from one type of system to another, the Presentation Layer translates the different character codes so that the systems may communicate. The Presentation Layer encrypts, decrypts, and authenticates the data to prevent unauthorized access to information and to confirm the source of the information. In addition, the Presentation Layer also compresses data passed by the Application Layer to save space within the channel during transmission. At the receiving end, the Presentation Layer decompresses the data.

Application Layer

The Application Layer contains network applications and passes unmodified messages from the application to the Presentation Layer as a method for accessing services for the other OSI protocols.

bling, wireless radio signals, and satellite transmissions. Even though copper cabling still stands as the most common data-networking medium for LANs and the local loop, fiber-optic cabling supports wide-area networking and some local-area networking needs. The transmission of data through the use of radio frequencies or infrared signals has also begun to increase.

Before communication over a network can begin, data must encode as either an electrical or electromagnetic signal. With a local-area network, the Physical Layer has digital signal transmission and relatively short cable lengths. Rather than use Non Return to Zero Level, or NRZ-L, digital signals, local-area networks rely on Manchester encoding. Shown in figure 3.2, NRZ-L has limitations that prevent its use for communication. Moving to figure 3.3, Manchester encoding indicates a "one" through a high/low transition in the middle of a bit and a "zero" through a low/high transition in the middle of a bit. When compared to the NRZ-L digital signal, Manchester encoding has no dc component and includes synchronizing information with the data signal.

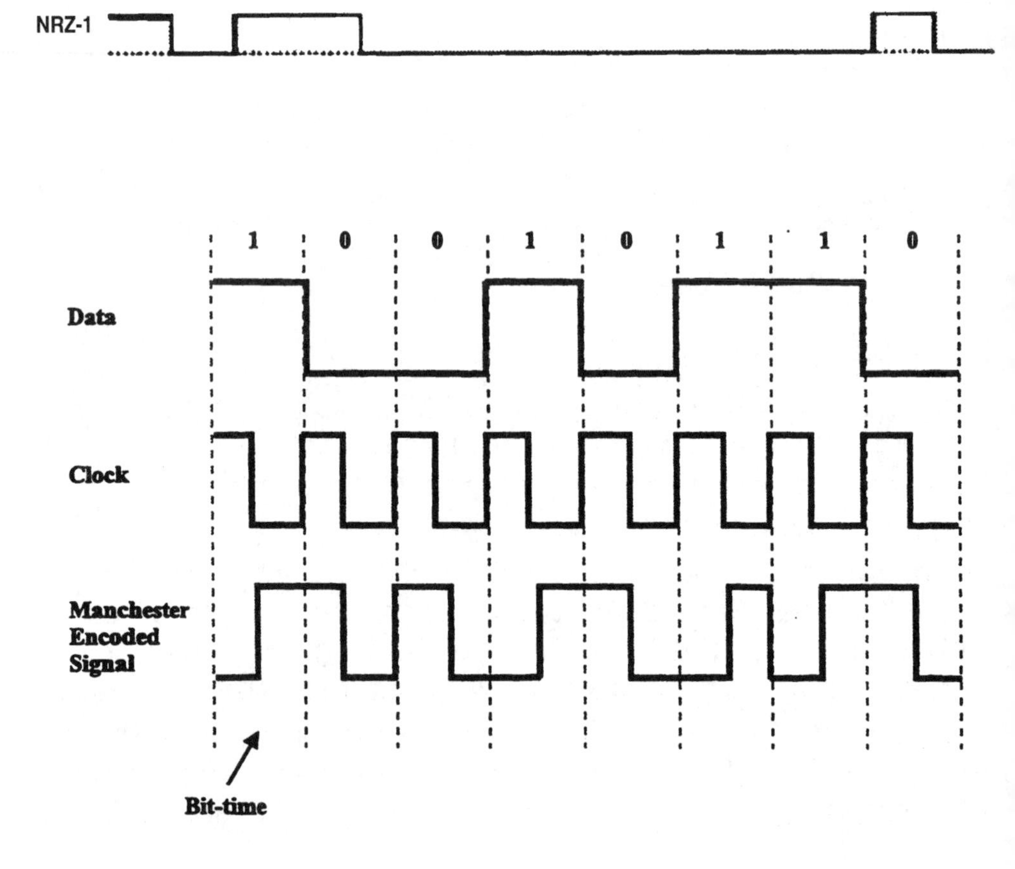

Figure 3.2. NRZ-1 digital signal format

Figure 3.3. Manchester encoding

The Subnet Protocol

Within the Physical Layer, the subnet — or transportation — protocol varies according to the type of physical link used within the network. The subnet protocol establishes the method for representing bits of data. In addition, the subnet protocol also provides a method for showing when a transmission begins and ends. Along with those functions, the subnet protocol indicates whether bits can flow in only one direction or in both directions simultaneously.

All this movement depends on the type of physical link used within the network. As an example, the subnet protocol for twisted-pair cabling has a different set of standards than the subnet protocol used for fiber-optics cabling. Examples of subnet protocols include the Ethernet and Token Ring networking standards as well as the RS-232 transmission standard.

The Physical Layer and High-Speed Connections

The OSI Physical Layer supports high-speed data transmission through the application of three sublayers. While the Physical Coding Sublayer, or PCS, encodes data, the Physical Medium Attachment Sublayer, or PMA, maps messages from the PCS to the transmission media. The Medium-Independent Interface, or MDI, specifies the type of connector for the application. Gigabit Ethernet networks also rely on the Gigabit Medium Independent Interface, or GMII, for the support of different decoding and encoding methods.

Applications for the Physical Layer

All cables, connections, and equipment such as hubs, routers, and switches occupy the Physical Layer. The specifications and categories that cover cabling, relays, repeaters, and cabling interfaces fit the requirements of the OSI model. As an example, the twisted-pair cabling used for local-area networks offers certain bandwidth and data-carrying capabilities. Fiber-optic cables also fit into the Physical Layer but have greater bandwidth and data-carrying capabilities.

Cables and Connectors

Twisted-Pair Cabling

Almost every building in America equipped for telephone services uses twisted-pair cabling to carry the telephone and other communications signals. Because the signals have become more complex and because more sources of interference have surfaced, the twisted-pair cabling industry has experienced change. Even with the application of fiber-optic cabling, twisted-pair cabling must now have the capability to carry high data-rate signals and to reject noise interference from industrial and telecommunications sources such as electric motors, power lines, and high-power radio and radar signals.

Twisted-pair cabling begins with two copper wires encased in color-coded insulation and twisted together to form one twisted pair. Then, the manufacturer packages multiple twisted pairs in an outer jacket to make the twisted pair cable. Varying the length of the twists minimizes the possibility of interference between pairs packaged in the jacket.

Twisted-pair cabling used for telephone and local-area network applications provides limited bandwidth and work with the baseband transmission of signals. Shown in figure 3.4, unshielded twisted-pair, or UTP, cabling has become one of the most popular types of transmission media and originated with the cabling used for telephone connections. Because UTP combines low cost, ease of installation, flexibility, and the capability for carrying relatively high data rates, data commu-

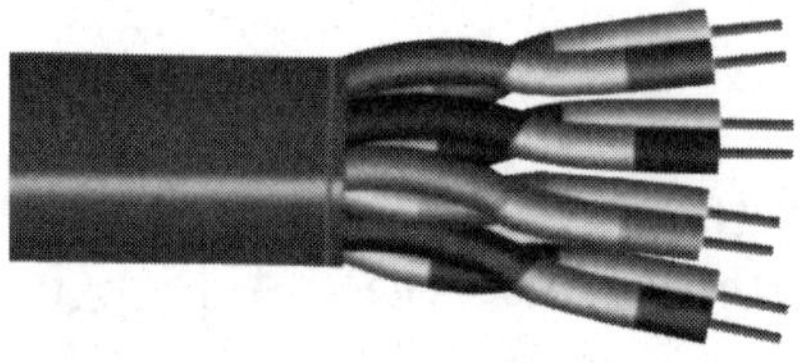

Figure 3.4. Unshielded twisted-pair cable

nications and telecommunications installations continue to use the unshielded twisted-pair standard for millions of network connections.

To obtain optimal performance, UTP cable should work as part of an well-engineered structured cabling system. As an example, the installation of UTP cabling requires the use of balance transformers called baluns, or the use of media filters. In either case, the equal induction of noise into the two conductors cancels the noise out at the receiver.

Shielded twisted-pair, or STP, cable provides additional protection against noise interference and includes screened twisted-pair cable and foil twisted-pair cable. Even though some similarities occur, STP cable has a slightly different design and manufacturing process than UTP cable. STP cable encases the signal-carrying wires within two shields.

The shields act as an antenna and convert received noise into current flowing along the shield. In turn, the current induces an equal and opposite current flowing in the twisted pairs. As long as the two currents remain symmetrical, the currents cancel one another and deliver no net noise to the receiver. Any break in the shield or difference between the quantity of current flowing along the shield and the quantity of the current flowing in the twisted pairs establishes noise.

As a result, the effectiveness of STP cabling at preventing radiation or blocking interference depends on the proper shielding and grounding of the entire end-to-end link. The effectiveness of the shielding becomes more important because of the characteristics of STP cable. For example, the attenuation of STP cabling may increase at high frequencies. In addition, crosstalk and signal noise may increase without compensation for the shield.

Shield effectiveness depends on the shield material and thickness, the type and frequency of the electromagnetic interference, the distance from the noise source to the shield, the continuity of the shield, and the grounding structure. For some applications, STP cables use a thick braided shield that increases the weight, thickness, and difficulty of installation for the cable. Other STP cables — called screened twisted-pair, or ScTP, cabling and foiled twisted-pair, or FTP, cabling rely on a relatively thin overall outer foil shield and have a decreased thickness and cost. In addition, the ScTP and FTP cabling has a maximum pulling tension force intended to prevent the tearing of the shield.

STP cabling consists of four 24-gauge cables that, in turn, consist of 100 twisted pairs. As mentioned, ScTP and FTP cabling wrap a foil shield around the cable package. Fully shielded twisted-pair cabling encloses each of the four cables with an individual foil shield. An overall shield complements the individual shield.

The dependence on proper shielding also carries over to the installation of connectors and other hardware for STP cable. Improperly shielded connectors, connecting hardware, or outlets can cause the degradation of the overall signal quality and noise immunity. A well-installed shielded cabling system requires the full and seamless shielding of every component within the system along with good grounding practices.

Twisted-Pair Categories

Because of changing application requirements, UTP and STP cabling has become available in different specifications called categories. Although basic telephone cable — or direct-inside wire — continues to provide value for voice connections, improvements in the manufacture of UTP cabling include variations in the twists, individual wire sheaths, or overall cable jackets. The application requirements and the capability to manufacture improved cabling have led to the development of the three EIA/TIA-568 standard-compliant categories shown in table 3.1.

Category	Maximum Signal Bandwidth
Cat 1	Up to 1 MHz(Used only for analog voice transmission)
Cat 2	Up to 4 MHz (Used for low-speed IBM Token Ring networks)
Cat 3	Up to 16 MHz (widely-used for digital voice transmission and for some 10BaseT Ethernet networks)
Cat 4	Up to 20 MHz (has few existing uses)
Cat 5 Cat 5e (Category 5 Extended)	Up to 100 MHz (used for a wide-range of 10BaseT, 100BaseT, and 1000BaseT networking applications)
Cat 6	Up to 250 MHz (intended for high-speed network applications)
Cat 7	Up to 600 MHz (intended for high-speed network applications)

TABLE 3.1 — TWISTED-PAIR CABLING CATEGORIES AND SIGNAL BANDWIDTH SPECIFICATIONS

Note: TIA standards use the term "category" to specify both components and cabling performance. ISO/IEC standards use the term "category" to describe component performance (i.e., cable and connecting hardware). The term "class" describes cabling characteristics such as link and channel performance.

Categories 3 and 4 Twisted-Pair Cabling

As table 3.1 shows, the six categories of transmission performance specified for cables, connecting hardware, and links are Cat 3 through Cat 7. Category 3 cabling transmission requirements specify the capability to carry an upper frequency limit up of 16 MHz. For category 4 cabling, the upper frequency limit moves up to 20 MHz. Because of the similarities between Cat 3 and Cat 4 cabling and the introduction of other cable standards, category 4 cabling has remained virtually unused.

Category 5 Twisted-Pair Cabling

Category 5 cabling has a larger diameter and, as a result, higher bandwidth capabilities than smaller gauge cables. Referring to table 3.1, note that Category 5 cabling has an upper frequency limit of 100 MHz. The newer superset of Category 5, called extended category 5 or Cat 5E, also has an upper frequency limit of 100 MHz.

Because Category 5 cabling has become the standard for networking applications, some variations of the standard have exhibited varying degrees of performance. Jacketing material, conductor insulation, and physical shape can cause performance inconsistencies among Category 5 cable products from different manufacturers. As a result, a wide range of inconsistencies may exist within the Category 5 designation.

To counter these problems, manufacturers have used the "meets or exceeds Category 5 standard" designation to show that cabling surpasses the standards set for Category 5 cabling by a factor of 2. Attenuation-to-crosstalk ratio has become the benchmark for evaluating any product called Category 5 cabling.

Under actual operating conditions, Cat 5 cable becomes severely attenuated at frequencies above 100 MHz. As the signal frequency increases, attenuation and NEXT increase because the electromagnetic field that interferes with signals on adjacent wires builds as the frequency increases. In comparison, the attenuation-to-crosstalk ratio decreases with frequencies of 115 MHz and above. NEXT levels exceed the signal level and cause the ACR to change from a positive to a negative value.

Categories 6 and 7 Twisted-Pair Cabling

Another new standard — Category 6 — establishes the best performance range delivered by unshielded and screened twisted-pair cabling. Category 6/class E cabling has an upper frequency range of 250 MHz and requires an eight-position modular jack interface at the work area. In addition, Category 6/class E components and cabling will offer backward compatibility for applications running on lower categories and classes.

The new Category 7/class F describes a new performance range for fully shielded twisted-pair cabling and has a bandwidth of 600 MHz. When introduced, Category 7/class F cabling and components require a new modular plug and socket design and offer the same backward compatibility seen with Category 6/class E cabling and components.

Coaxial Cable

Pictured in figure 3.5, coaxial cable has a solid center conductor that is surrounded by an insulating spacer. In turn, a tubular outer conductor consisting of either braided or foiled material surrounds the inner conductor and insulator. Another outer layer covers the entire cable assembly and provides both insulation and protection. Coaxial cables work well for data transmission because of a wide bandwidth and the capability to carry multiple data, voice, and video streams simultaneously.

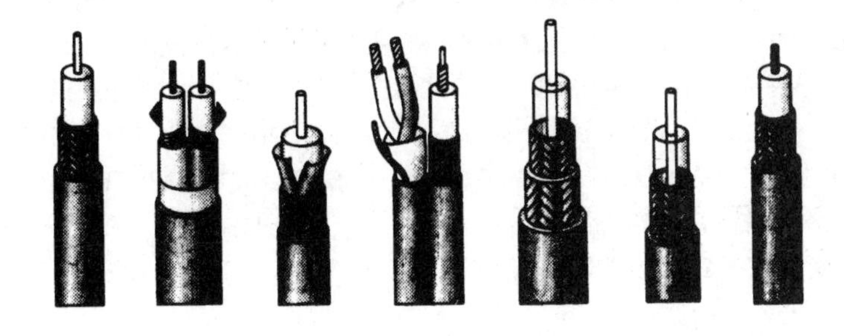

Figure 3.5. Coaxial cables

Thinnet

Some network applications require a type of coaxial cable called Thinnet, which provides a 50-ohm impedance and a small 5-millimeter diameter. Thinnet supports a data transmission rate of 10 Mbps at a maximum segment length of 185 meters. Because of the smaller diameter, Thinnet offers low cost, lightweight, flexibility, and easy installation. However, Thinnet has the drawback of transmission characteristics that support a shorter maximum segment length and fewer connected nodes per segment than the thicker coaxial cable types.

Thicknet

Compared to Thinnet, Thicknet offers longer segment lengths and the capability to support more nodes per segment. When used for 10Base-5 applications, Thicknet supports maximum segment lengths of 500 feet and 50 nodes per second. As with Thinnet, a Thicknet coaxial cable has a 50-ohm impedance. However, as the name implies, the cable has a larger diameter at 10 millimeters. Thicknet Ethernet cables have bright outer jack colors punctuated with black bands placed at 2.5-meter intervals to mark the installation points for transceivers.

Twinax

Another type of coaxial cable called Twinax, or twinaxial, consists of two center conductors surrounded by an insulating spacer. As with the other coaxial cable types, an outer conductor surrounds the inner conductors and insulation. Then, another layer of insulation covers the entire assembly. Given its construction, Twinax offers the durability of coaxial cable and the balanced transmission characteristics of twisted-pair cabling. The 150-ohm Twinax works as a "short haul" cable with the 1000Base-CX media system. Although Twinax has better transmission characteristics than twisted-pair media, it supports segment lengths of only 25 meters for 1000Base-CX because of the very high 1.25-Gbps data transmission rate seen with that standard.

Fiber-Optic Cables

Many building-to-building and WAN installations use fiber optics. With this, an insulator encloses a bundle of glass threads. Each thread can transmit information at almost the speed of light. Along with the advantage of speed, fiber optics also offers benefits such as:

- Greater bandwidth,
- Less susceptibility to interference,
- The ability to carry digital signals, and
- Reduced size and weight.

Along with those basic advantages, optical fiber also offers several specific benefits not seen with copper media. In particular, the use of fiber-optic cabling enhances network security, reliability, and performance because fiber does not emit electrical signals. In comparison, copper cabling emits electrical signals and remains susceptible to tapping and the unauthorized access to carried data. In addition, fiber-optic cabling offers immunity to radio frequency interference, or RFI, and electromagnetic interference, or EMI.

Until recently, the use of fiber optics had not become widespread due to the installation cost and the fragile characteristics of the medium. Despite those factors, fiber optics has become more popular for LANs and telephone networks. Nearly every telephone company in the nation has replaced or plans to replace existing copper lines with fiber optics. Table 3.2 displays the differences between the media types.

Transmission Media	Data Rate	Cable Length
Unshielded Twisted Pair	1-100 Mbps (dependent on category)	0.1 km
Shielded Twisted Pair	16 Mbps	0.3 km
Coaxial Cable	70 Mbps	Greater than 1 km
Fiber Optics	100 Mbps	Greater than 1 km

TABLE 3.2 — TRANSMISSION TYPES, DATA RATES, AND MAXIMUM CABLE LENGTHS

Even though costs have decreased, the per-mile cost for laying fiber-optic cables continues to surpass the cost of laying copper lines. The cost savings seen with fiber-optic cables occur at the per-circuit level when used for long-distance applications. At shorter distances such as a local loop, the use of fiber-optic cables becomes impractical because of the low number of circuits. Figure 3.6 shows several types of fiber-optic cables, and figure 3.7 shows a selection of fiber-optic cable connectors.

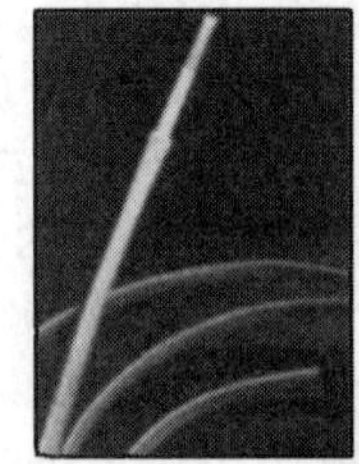
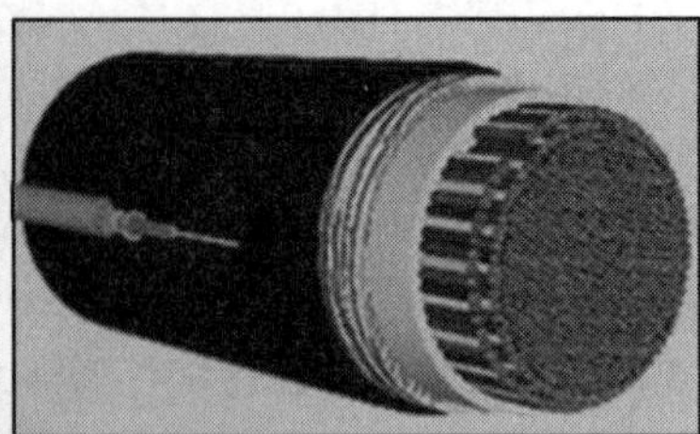
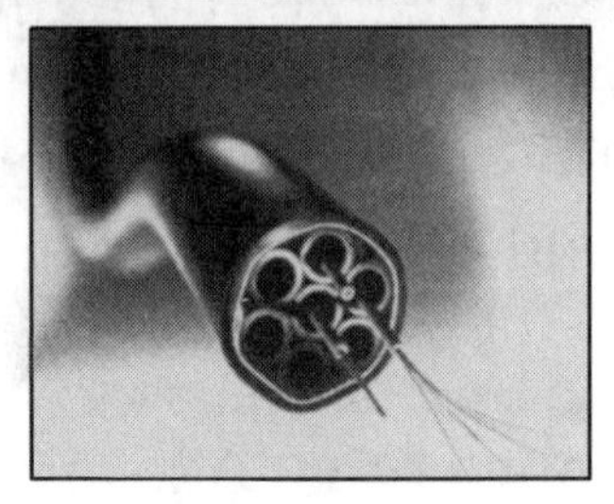

Figure 3.6. Fiber-optic cables

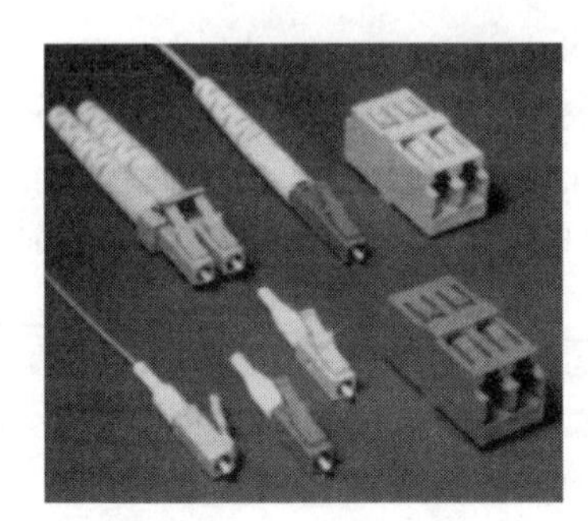

Figure 3.7. Fiber-optic connectors

Single-Mode and Multimode Fiber-Optic Cables

Manufacturers produce fiber-optic cables according to mode specifications. We can define a mode as a ray of light that enters the fiber at a specific angle. In terms of communications, two types of fiber-optic cabling — multimode and single-mode — receive the most attention. Both types have a total strand diameter of about 125 micrometers. However, multimode fiber-optic cable has a larger core transmission medium than single-mode fiber-optic cable.

Applications for multimode fiber allow multiple modes of light — produced by Light Emitting Diodes, or LED — to propagate throughout the fiber. With each mode entering the fiber at a different angle, the modes arrive at the destination at different times and become part of a characteristic called modal dispersion. Because of modal dispersion, the bandwidth capability and transmission distances with multimode fiber-optic cabling remain limited. As a result, applications for multimode fiber provide connectivity within a building or a small geographic area.

As shown in figure 3.13, a laser beam provides the single-mode fiber. The thinner core transmission medium found in the single-mode fiber-optic cable allows only one mode to travel down the strand. As a result, modal dispersion does not occur. In addition, regenerating the signal at points along the span becomes easier, and the cable can carry data at a faster rate than seen with the multimode type. Single-mode fiber-optic cables have become the standard for long distance networks

Connectors

Four basic modular jack styles exist in data communications. Rather than refer to the jacks as RJ-45 or a RJ-11 jack, the more correct method refers to the number of cable positions found in the jack. As an example, we refer to the jacks commonly used for UTP cabling as eight-position modular outlets. The letters "RJ" designate the jack as a registered jack and actually refer to specific wiring configurations called Universal Service Wiring Codes. A technician could wire each of the basic jacks discussed in this section for several different types of RJ designations. Figure 3.8 diagrams a RJ-45 connector.

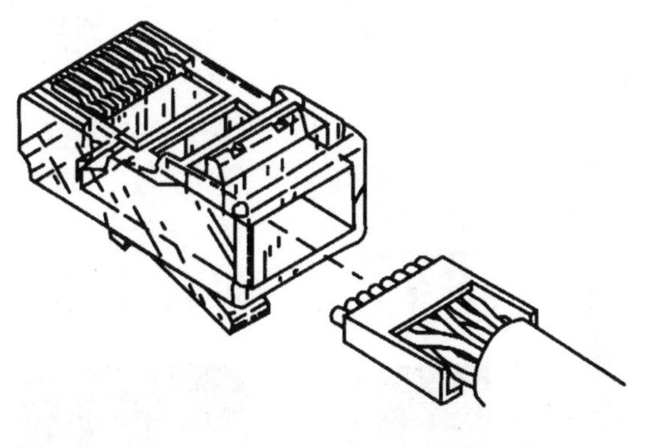

Figure 3.8. RJ-45 connector

One application for modular cables involves using the cables as "patches" between modular patch panels. For this application, the modular cables should always have a straight-through configuration. Another application for modular cabling involves the connection of a personal computer, telephone, FAX machine, or other electronic equipment to a modular jack.

Network Interface Cards

The physical connection to the network occurs through the placement of a network interface card, or NIC, inside the computer and connecting the card to the transmission cable. Figure 3.9A shows a network interface card, and figure 3.9B depicts the operation of the NIC through a block diagram. Because the expansion slot connects the NIC to the processor and memory of the computer, it allows the transfer of data to and from the computer to the network. Even though many computer systems rely on a separate network interface card, some manufacturers have begun to integrate the Ethernet controller functions onto the PC motherboard through the use of a single integrated circuit.

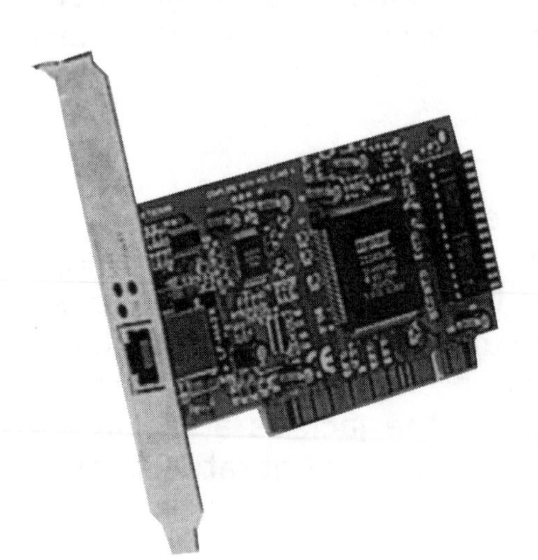

Figure 3.9A. Network interface card

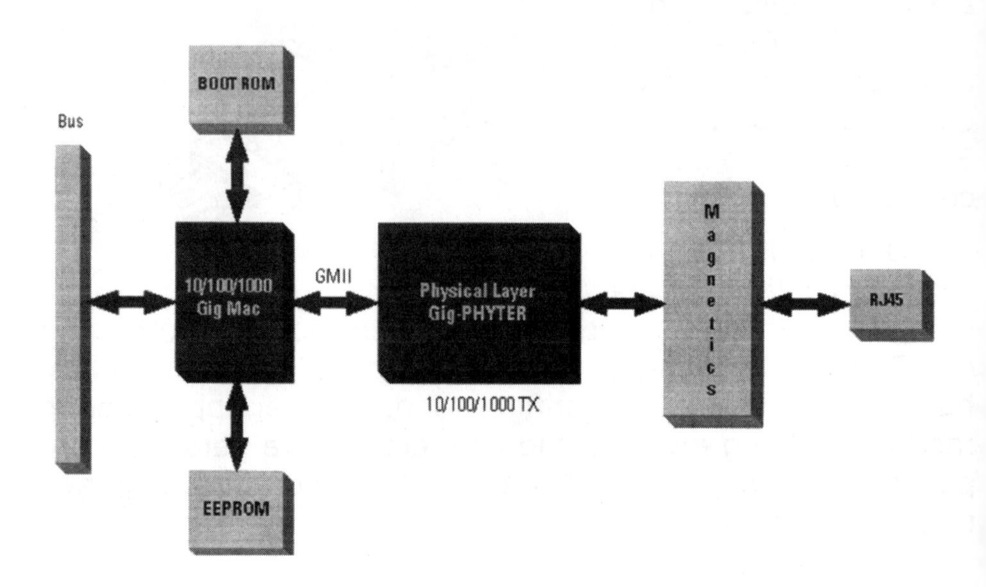

Figure 3.9B. Block diagram of a 10/100/1000 network interface card

Data transfer methods for a network interface card include programmed input/output transfers where the NIC stores data at the CPU for transfer to RAM. In addition, NICs transfer data by sharing the computer memory and interrupting the processor. Two other methods for data transfer involve bus mastering, or the transfer of data from the NIC to the system memory, and direct memory access where the NIC interrupts the system microprocessor from routine tasks. The microprocessor initiates the transfer to the system memory. Network interface cards arrive configured for the type of network, the speed of the network, and the type of computer bus.

Once the physical connection is in place, the network software manages communications between stations on the network and the server. Network interface cards support network operating systems such as Windows NT, Novell NetWare, LAN Manager, VINES, and AppleTalk. During the installation of the network operating system, the configuration of the software involves the selection and setup for the installed card.

Repeaters

A repeater functions at the physical layer and interconnects the media segments of an extended network. Repeaters receive signals from one network segment and amplify, retime, and retransmit those signals to another network segment. With this, the repeater allows a series of cable segments to act as a single cable. The amplification gained through the use of repeaters prevents signal deterioration caused by long cable lengths and large numbers of connected devices.

Figure 3.10 illustrates a functional diagram of a repeater. As the figure shows, repeaters do not perform complex filtering and other traffic processing. Moreover, the amplification given by a repeater applies not only to desired signals but also to noise and network errors. Because of these factors and the possibility of timing errors, most network designs limit the usage of repeaters.

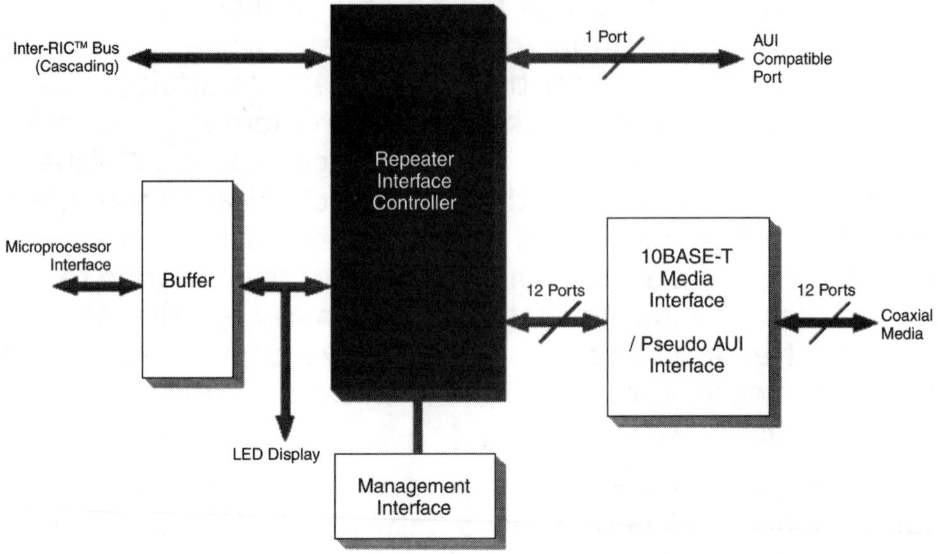

Figure 3.10. Block diagram of a repeater

Extenders

An extender, pictured in figure 3.11, operates as a remote-access multilayer switch that connects to a host router. Extenders forward traffic from all the standard network-layer protocols and filter traffic based on the MAC address or network-layer protocol type. However, extenders do not segment LAN traffic or create security firewalls.

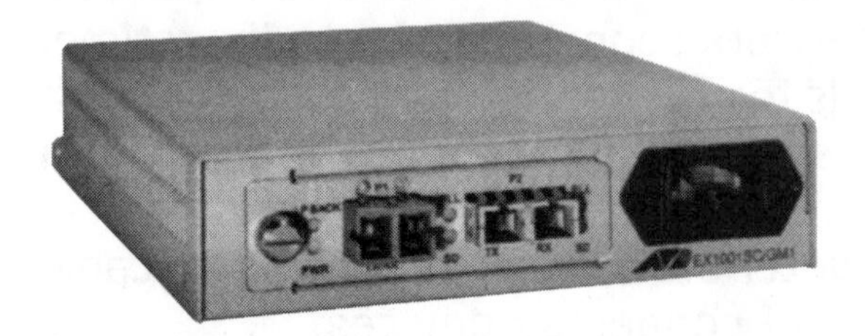

Figure 3.11. Extender

Hubs

Referring to figure 3.12, a hub connects multiple user stations through dedicated cables and establishes electrical interconnections. With this, the hub operates much like a multiport repeater. As shown in figure 3.13, a hub creates a

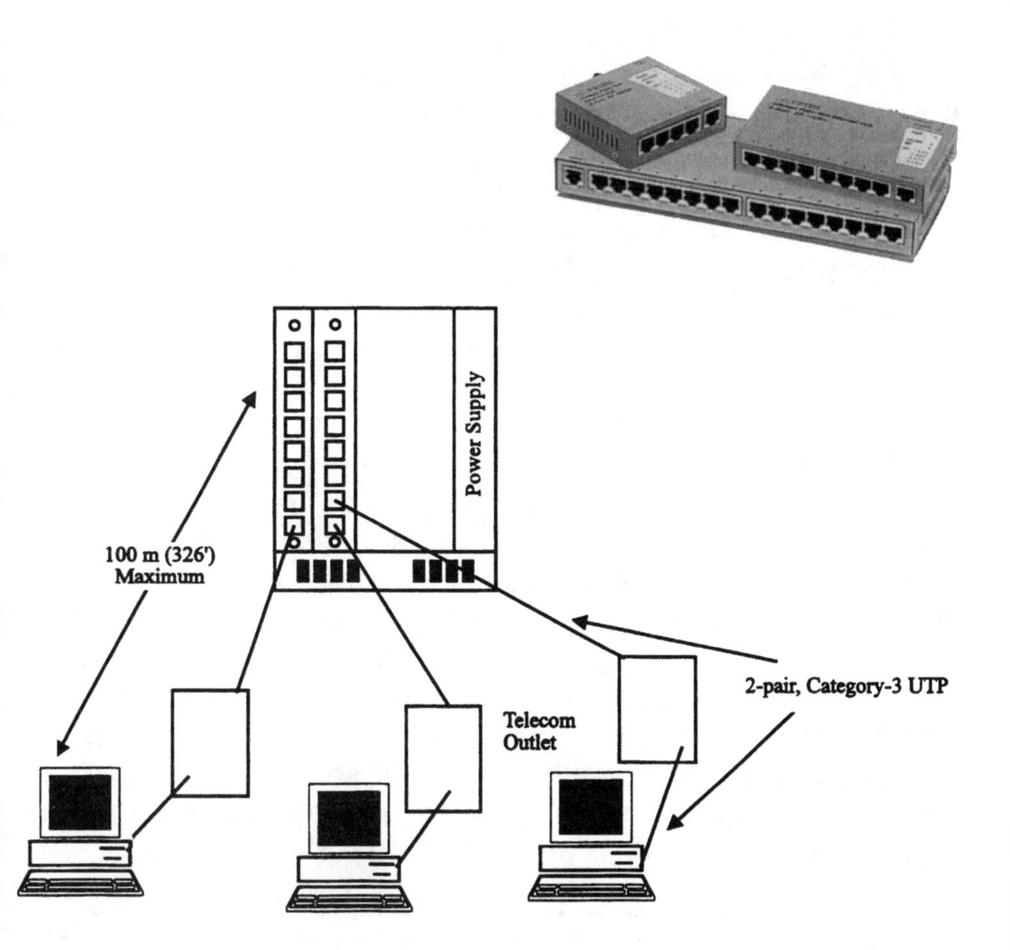

Figure 3.12. Hubs

Figure 3.13. Hub operates at the center of a physical star network

physical star network while maintaining the logical bus or ring configuration of the LAN.

Passive hubs concentrate multiple connections into a single device and use electrical relays to connect all users. In some network installations, a number of passive hubs may connect together in a daisy chain arrangement. Intelligent hubs add management capabilities to the connections provided by the passive hubs. With this, the hub provides status reports about the connections, compiles statistics about connection usage, and supports the connection of Ethernet and Token Ring cards.

- Hubs are installed in wiring closets centrally located in a building.
- Hubs detect when a node is not responding and "lock it out" so that the ring can continue to operate when a node fails. This happens automatically when the hub senses that a node is not responding.
- Hubs provide a "bridge" to other rings. With this connection, the hub sends messages addressed to nodes on other rings across the bridge circuits to those rings and accepts messages from other rings for its nodes.

Stackable Hubs

A stackable hub consists of hubs that stack on top of one another and interconnect through short cables. Even though separate hubs make up the stackable hub, the network "sees" the hub as a single logical device. Stackable hubs offer increased network management capabilities through the use of managed and manageable hubs, flexibility through the availability of more ports, and lower prices per port. The economy and benefits given through the use of stackable hubs occurs because the packets traveling between the devices do not depend on the electronics contained within the repeater.

The repeater units in the hubs connect and form a single device. Each manufacturer of stackable hubs implements the intrahub connections in different ways. In all instances, the patch cables connect repeaters found within the hubs and carry the network data and control information. However, hubs working within the stack must come from the same manufacturer. Packets moving between the hubs do not depend on the equipment type.

Concentrators

One solution for expanding an Ethernet network involves the cascading, or connecting, of hubs to one another. The cascading of four eight-port hubs can expand the network from the original 8 devices to 32 devices. Moving to figure 3.14, a

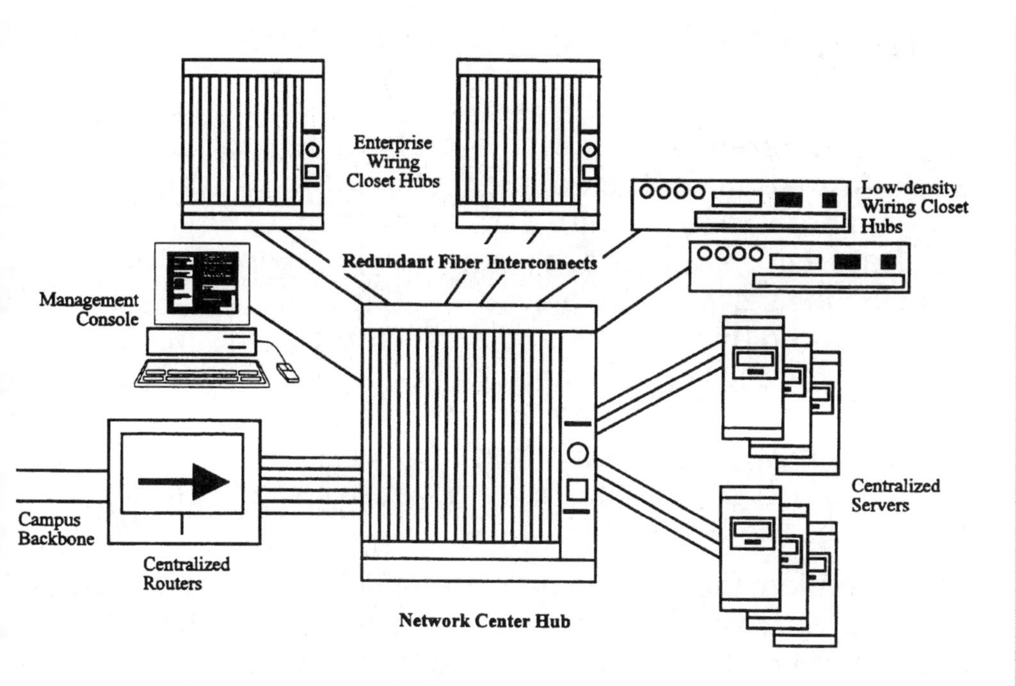

Figure 3.14. Structured network system using hubs

special type of hub called a concentrator eliminates the need to install cascaded hubs as network requirements grow. Instead, the concentrator consists of a chassis that accepts add-on module boards. Because the backplane of the concentrator serves as a cascaded cable, the installation of the modules easily increases the number of available ports.

Intelligent Hubs

An intelligent hub incorporates a microprocessor that can perform preprogrammed functions. The functions may include

- recognizing commands for turning ports on and off,
- obtaining and displaying usage statistics,
- displaying diagnostic information, and
- allowing the viewing of network traffic from a console.

Although intelligent hubs enhance the management capabilities of the network, the hub continues to operate as a shared media device.

The Data Link Layer

The different layers of the OSI model work with data and information contained in different units. At the Physical Layer, a network transmits data in the form of bits, and then the Data Link Layer moves data in the form of frames. A frame consists of data placed in a specific sequence and encapsulated by a header and a trailer that contain addressing and messaging information. At the Network Layer, networks transmit datagrams or packets as units of information that may consist of a message or a segment of information. Each packet includes a header that contains the address of the destination node.

Again looking at the complete OSI model, the Data Link Layer covers the transmission of frames rather than the transmission of bits of data as seen with the Physical Layer. Given the purpose of the Data Link Layer, it functions in tandem with the Physical Layer and the Network Layer. Within this function, the Data Link Layer may provide either connectionless transmission that requires no acknowledgment or connection-oriented transmission that requires an acknowledgement of service.

An access method defines the set of rules used within networks to arbitrate the use of a common medium. Without access methods, different streams of data would collide as users share a network. All access methods operate at the Data Link Layer and cover the efficient movement of data rather than content. Examples of access methods include Ethernet and Token Ring.

Connectionless Services

Connectionless transmission works well if the Data Link Layer provides high reliability and the upper layers of the OSI model provide error correction. With connectionless service, a protocol designates each individual data packet as an individual unit, and the network ensures that the packet reaches the proper destination. Because each packet exists as an individual unit, the network does not establish and maintain a connection before sending and receiving the packet. During connectionless operation, the Data Link Layer breaks the stream of data received from the upper layers into frames. Referring to the Data Link Layer only, connectionless operation does not have a great amount of complexity.

Connection-Oriented Networks

Connection-oriented network operation uses a protocol where the network establishes and maintains a connection through out the data transaction. The stations involved in the transaction remain in contact with one another during the sending and receiving of the packets. During connection-oriented operation, the Data Link Layer

- establishes a connection,
- accepts packets of data from the Network Layer,
- divides the packets into frames, and
- passes the frames to the Physical Layer for transmission.

Within this framework, the Data Link Layer becomes active at both the transmitting and the receiving ends of the network.

At the transmitting end of the network, the Data Link Layer accepts the address of the node that contains a Physical Layer link so that the network can transmit data to the node. Then, the Data Link Layer accepts data packets from the Network Layer. As soon as the Data Link Layer accepts the packets, it begins to handshake with the peer to ensure the correct reception of the data. Handshaking is the initial setup process that occurs between two computers attached to a network and establishes the communication parameters between the two nodes.

Handshaking and the Data Link Layer

At the receiving end of the network, handshaking begins the process of passing data packets back to the Network Layer. During this process, the Data Link Layer ensures that each frame retains the proper sequencing information. In addition, the Data Link Layer applies error detection and correction codes to the frames. With this, the receiving process can generate a report to show the location of an error.

At the receiving end of the transmission, the handshaking information added by the Data Link Layer works with the peer process to correct problems that may occur. Examples of network errors include corrupted, lost, delayed, and duplicate data. In addition, the data may arrive at the destination node out of sequence. The Data Link Layer protects against errors by regulating the speed of the transmission and by requesting that the transmitting peer retransmit any data that shows errors. All this combines with the error detection and correction offered in higher layers of the OSI model.

Because the Data Link Layer defines data formats, the error detection, location, and correction functions found within the Data Link Layer may exist as the most important functions provided by this layer. In brief, the Data Link Layer has the responsibility for the reliable delivery of information. Data Link Layer protocols include the Binary Synchronous Communications, or BSC, protocol and the High-Level Data Link Control, or HDLC, protocol.

Subdividing the Data Link Layer

The original OSI model served as a standard for wide-area networking. With the implementation and popularity of local-area networks, the need for modifications to portions of the original model became apparent. While the Logical Link Control, or LLC, generates and interprets commands that control the flow of data, the Media Access Control, or MAC, sublayer provides access to the LAN. After the development of high-speed local-area networks, engineers added the Reconciliation Layer, or RL, as a method for connecting medium and high-speed interfaces.

Logical Link Control

The IEEE 802.2 standard defines the Logical Link Control sublayer as the method of link control that remains independent of any specific access method. With this independence, the LLC generates and interprets commands to control the flow of data across Ethernet, token bus, and Token Ring local-area networks. In addition, the LLC provides recovery for transmission errors.

As figure 3.15 indicates, the data field of an 802.3 frame carries the LLC information as an LLC protocol data unit. Within the OSI model, the Logical Link Control sublayer fits between the Medium Access Control sublayer also contained within the Data Link Layer and the Network Layer. Service Access Points pass information between the Network Layer and the two sublayers.

Header (size is variable)	Destination Address (8 bits)	Source Address (8 bits)	Control Field (8 or 16 bits)	Data (8 x n bits)	Trailer (size is variable)

Figure 3.15. Logical Link Control Layer frame

Three types of service allow the sending and receiving of LLC data. The Type 1 unacknowledged connectionless service allows the transmission to flow on the channels to all stations, whereas the Type 2 connection-oriented service requires the establishment of a logical link between the sending and receiving stations. Data travels only between the two stations rather than on the channels of all stations. The Type 3 acknowledged connectionless service provides for setting up and disconnecting the transmission by acknowledging individual frames with flow control.

Although the Type 1 service does not feature the acknowledgment of frames, flow control, or error recovery, all logical link control stations support Type 1 operations. Most protocol suites rely on a transport mechanism within the transport layer of the OSI model and do not require the listed reliability features at the Data Link Layer. Because of the lack of overhead, the Type 1 service also provides greater throughput than the other services.

Medium Access Control

The Medium Access Control sublayer checks either an active channel for the status of the transmitting data or an inactive channel for the occurrence of a collision. If the MAC sublayer detects a collision, it performs a series of predefined steps and provides the necessary logic to control the network. In addition, the MAC sublayer serves as an interface between user data and the physical placement and retrieval of data on the network.

The functions of the MAC sublayer separate into the transmitting and receiving of data operations and the transmitting and receiving of medium access management. With the transmission of data operations, the MAC sublayer accepts data from the LLC sublayer. Then, the MAC sublayer constructs a frame by appending the preamble and start-of-frame delimiter and then inserting the destination and source address. If the frame has a length of less than 64 bytes, the MAC sublayer inserts PAD characters into the data field.

When the MAC sublayer receives data operations, it discards all frames not addressed to the receiving station and recognizes all broadcast frames and frames addressed specifically to the station. In addition, the MAC sublayer performs acyclic redundancy check, or CRC. The reception of data operations also involves the removal of the frame information placed during the transmission of data operations. At the end of the receive data operations process, the MAC sublayer passes the data to the LLC sublayer.

The transmission of medium access management presents a serial bit stream to the physical layer for transmission, checks whether any transmission of data has occurred on the network, and stops the transmission if a collision occurs. Detection of the transmitted data becomes possible through capability to monitor the Manchester encoding used on Ethernet local-area networks. After halting the transmission, the MAC sublayer transmits a signal that ensures that all stations receive notification about the collision. From that point, the sublayer reschedules the transmission after ensuring success or after a specified number of attempts has occurred.

The reception of media access management involves receiving a serial bit stream from the physical layer. In addition, the MAC sublayer verifies the byte boundary and length of frame. After inspecting the length of frame, the sublayer discards any frames that do not have a length of an even eight bits or a length of less than the minimum frame length.

Applications for the Data Link Layer

Regardless of the network type, the Data Link Layer serves the primary medium for the network. In comparison to the protocols seen with the upper layers of the model, the Ethernet, Token Ring, EtherTalk, and TokenTalk protocols provide the rules for transporting data across the network and — more specifically — the Physical Layer. In practice, the Data Link Layer also affects the connection of different medium. As an example, a computer that includes a Token Ring network interface card will not attach to an Ethernet network. Within Ethernet, Token Ring, and AppleTalk networks, different types of network cabling and styles of network interface card connections exist. However, all have the same purpose as set by the Data Link Layer standards.

Bridges

Available during the early 1980s, bridges originally connected and enabled packet forwarding only between the same type of networks. As technologies have improved, bridges have performed the same tasks between different types of networks. Today, bridges establish LAN-to-LAN connections for similar types of local-area networks. A bridge reads transmitted packets and determines the destination network through a process called filtering.

Bridges interconnect at the Media Access Control sublayer and route data packets using the Logical Link Control sublayer. Although a bridge cannot perform protocol conversion, it can interconnect LANs that rely on different communications protocols. With this, a bridge monitors all traffic on the subnets that it links and reads every packet. During this process, the bridge looks only for the MAC-sublayer source and destination address for the packet. Therefore, the bridge determines which subnet originated the packet and which subnet should receive the packet. Bridges provide these functions through the use of various Data Link Layer protocols that dictate specific flow control, error handling, addressing, and media-access algorithms.

Figure 3.16 shows the building of a packet containing application information by the host. The host also encapsulates the packet in a frame for transit over the medium to the bridge. At this point, the bridge strips the frame of the header at the MAC sublayer of the Link Layer. Then, the bridge passes the frame up to the LLC sublayer for further processing. After this processing, the packet passes back down for encapsulation into a header for transmission on the network to the next host.

Bridges provide filtering for frames based on any Layer 2 field or can reject unnecessary broadcast and multicast packets. As an example, a network administrator can program a bridge to reject all frames delivered from a particular net-

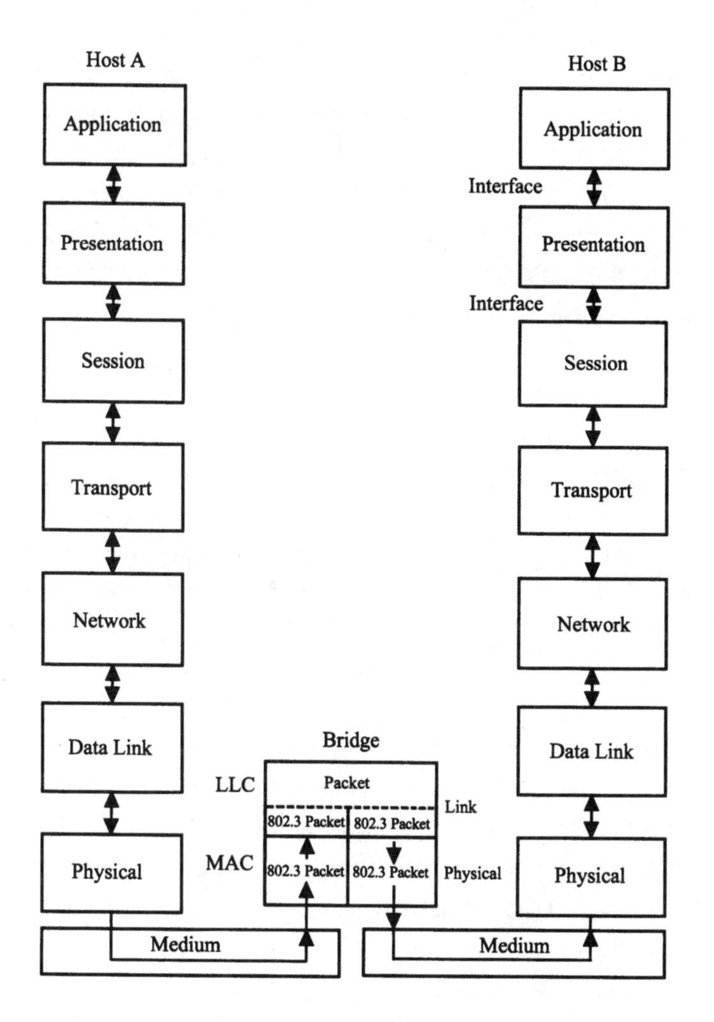

Figure 3.16. Building of a packet for transmission from host to host

work. Because of different types of network support for certain frame fields and the use of different protocol functions within different networks, a bridge may not produce perfect data translation.

Transparent Bridging

Connections between Ethernet-based local-area networks and wide-area networks have prompted the use of the transparent bridging scheme shown in figure 3.17. During network operation, a transparent bridge constantly updates address

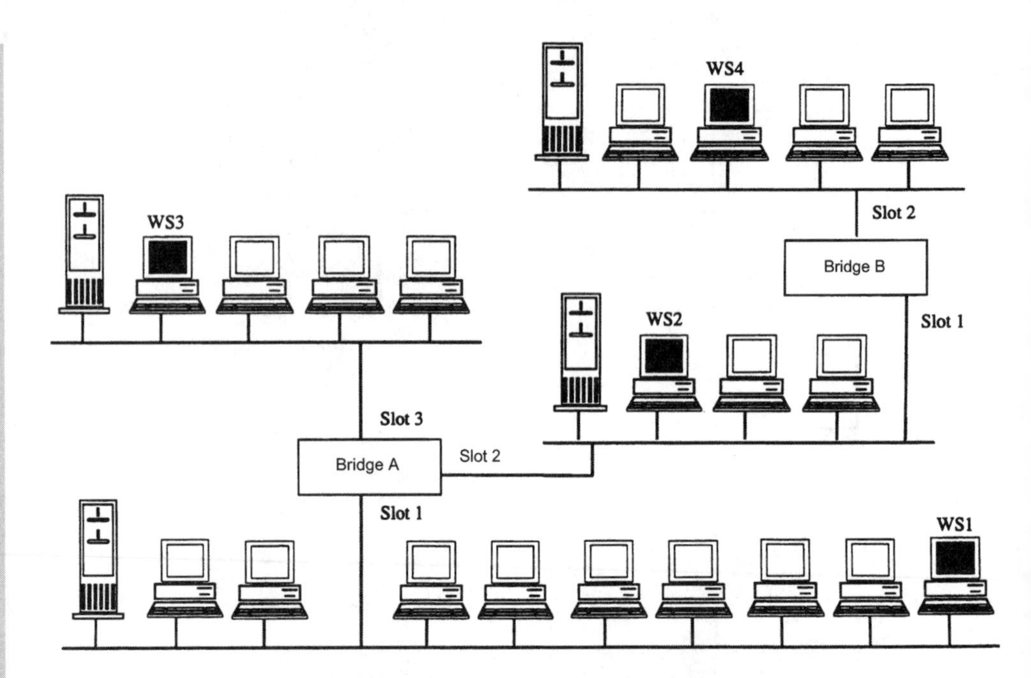

Figure 3.17. Transparent bridging

labels for the purpose of increasing performance. When a transparent bridge receives an information packet for a destination device not residing in its address table, it sends the packet through all its ports through a process called flooding. As soon as the bridge finds the destination address in the table, it forwards the packet.

Switches

Like bridges, switches enable the interconnection of multiple physical LAN segments into a single larger network and forward traffic based on MAC addresses. However, switches provide a significant improvement in transaction speed because the operation occurs within hardware rather than through software. Many types of switches exist for different applications such as local-area networking, ATM networks, and internetworking between LANs and WANs. Figure 3.18A shows a network switch, whereas figure 3.18B shows a functional diagram of the switch.

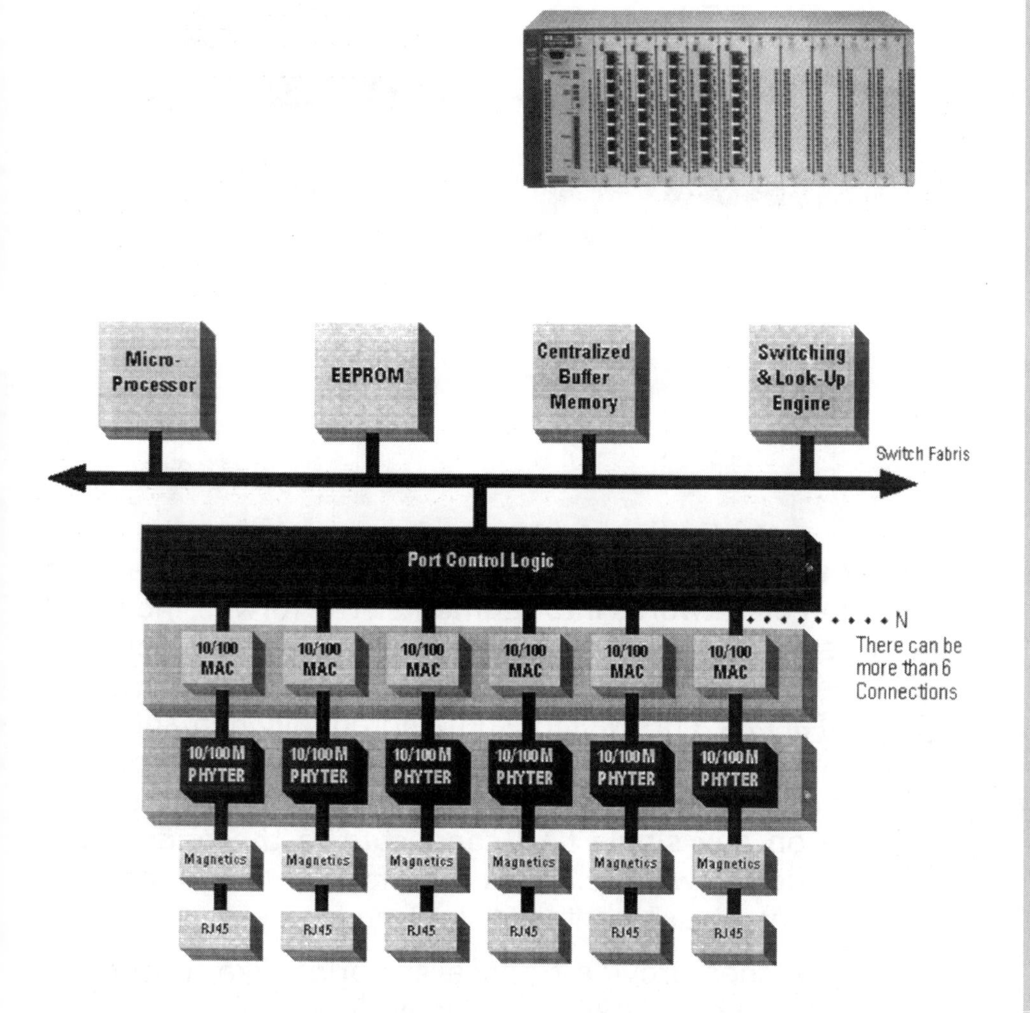

Figure 3.19a. A network switch

Figure 3.18B. Functional diagram of a switch

A LAN switch provides much higher port density at a lower cost than traditional bridges. For this reason, LAN switches can accommodate network designs featuring fewer users per segment. As a result, the use of a switch in a LAN increases the average available bandwidth per user. Figure 3.19 illustrates a simple network in which a LAN switch interconnects 10-Mbps and a 100-Mbps Ethernet LANs.

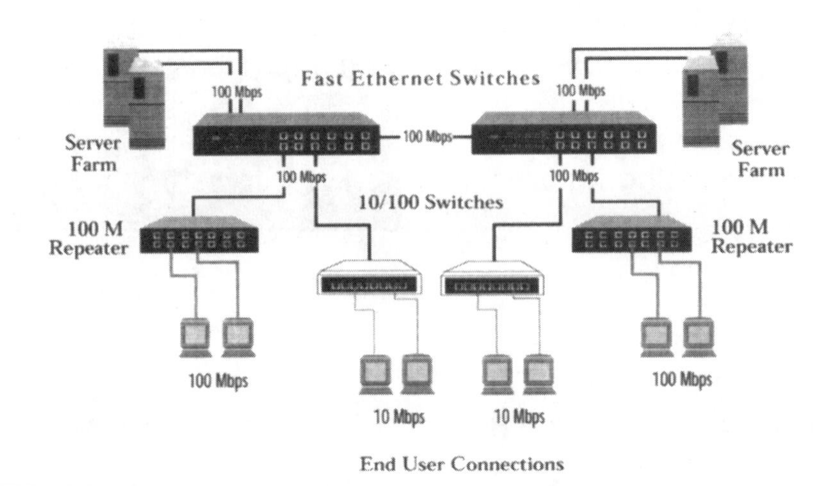

Figure 3.19. LAN switches interconnect 10- and 100-Mbps networks

Bridges and switches provide several advantages through the division of networks into segments. The forwarding of only a given percentage of traffic diminishes the traffic experienced by devices on all connected segments. In addition, a bridge or switch will act as a firewall for some potentially damaging network errors. Bridges and switches also accommodate communication between a larger number of devices than would be supported on any single LAN connected to the bridge. With this, a bridge or switch extends the effective length of a LAN and allows the attachment of distant nodes.

LAN switches provide transparent bridgelike functions in terms of learning a network topology, forwarding, and filtering. In addition, the switches also support dedicated communication between devices, multiple simultaneous conversation, full-duplex communication, and media-rate adaptation. Dedicated collision-free communication between network devices increases the speed of a file-transfer. Multiple simultaneous conversations can occur by either forwarding or switching several packets at the same time. As a result, network capacity increases by the number of supported conversations.

Full-duplex communication effectively doubles the data throughput. Media-rate adaptation refers to the ability of the switch to translate between 10- and 100-Mbps data rates. In addition, media rate adaptation allocates bandwidth on an as needed basis. With this, the installation of a switch requires no change to existing hubs, network interface cards, or cabling.

As we have seen, LAN switches interconnect multiple network segments and provide dedicated, collision-free communication between network devices while supporting multiple simultaneous conversations. In addition, LAN switches provide high-speed switching of data frames. To accomplish this, switches use either store-and-forward switching or cut-through switching when forwarding traffic.

The cut-through switching method reduces latency by eliminating error checking. From there, the switch looks up the destination address in its switching table, determines the outgoing interface, and forwards the frame toward its destination. The switch copies only the destination address — consisting of the first six bytes following the preamble — into its onboard buffers.

With the store-and-forward switching method, the switch copies the entire frame into its onboard buffers and computes the cyclic redundancy check. If the frame contains a CRC error, such as having fewer than 64 bytes or more than 1,518 bytes, the switch discards the frame. If the frame does not contain any errors, the switch looks up the destination address in its forwarding, or switching, table and determines the outgoing interface. Then, the switch forwards the frame toward its destination.

Some switches accept the configuration to perform cut-through switching on a per-port basis until the reaching of a user-defined error threshold. Then, the switch automatically changes to store-and-forward mode. When the error rate falls below the threshold, the port automatically changes back to store-and-forward mode.

Categories of Switches

Efficient network management requires the evaluation of the needed amount of bandwidth for connections between devices. With this, a network manager can ensure that the network accommodates the data flow of network-based applications. Much of the decision making in terms of switch selection depends on bandwidth requirements.

We can categorize LAN switches according to the proportion of bandwidth allocated to each port. Symmetric switching provides evenly distributed bandwidth to each port. During operation, a symmetric switch establishes switched connections between ports with the same bandwidth. As a result, symmetric switching remains optimized for a reasonably distributed traffic load.

Asymmetric switching provides unlike, or unequal, bandwidth between some ports. As opposed to the symmetric switch, an asymmetric LAN switch establishes switched connections between ports of unlike bandwidths. We can also refer to asymmetric switching as 10/100 switching. Asymmetric switching works well with client-server traffic in a network where multiple clients simultaneously communicate with a server. This type of traffic flow requires the dedication of more bandwidth to the server port so that a bottleneck will not occur at any particular port.

4

The OSI Model:

Network, Transport, Session, Presentation, and Application Layers

Introduction

Layers communicate through a basic form of messaging. Each message includes a header that identifies the type of message through control, size, and timing information. All messages receive a header; however, some also receive trailer information. In most cases, layer four of a network architecture will not have a size limit for the message, but layer three will establish a size limit. As a result, the transmission of messages from layer to layer may result in the reformatting of the message into packets. In turn, each packet receives a header.

Protocol Data Units, or PDUs, form the packets and carry peer process information used to perform peer protocols. Headers contained within the PDUs identify the PDUs that contain

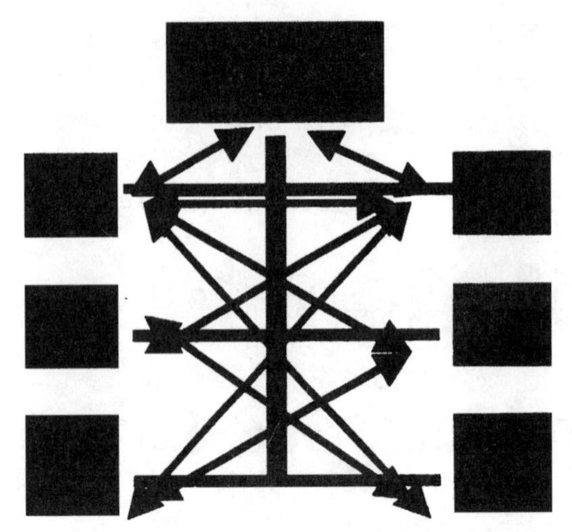

Figure 4.1. Diagram of peer processes in a network

data and PDUs that contain control information. The PDUs also provide the sequence and count information so that the receiving system layers can reassemble the fragmented data back into the original order.

As the packets pass from a lower layer to the next upper layer in the receiving computer, the upper layer strips away the corresponding header information before passing the packet to the next layer. Referring to figure 4.1, note that peer processes include horizontal transactions between layers. The lower layers of the network architecture implement through firmware-embedded hardware devices such as network interface cards or routers. Network software relies on the communication at the upper layers.

The Network Layer

Rather than sending frames across the network, the Network Layer sends packets of information. The Network Layer handles any problems associated with delivering a packet of information from one node of the network to another node. Although this process may seem simple, the network may include any number of nodes. The packet must travel through intermediate nodes because no direct connection exists between the source and destination nodes.

In Chapter 3, we found that the Data Link Layer only communicates with the peer process found at the end of the communications link. A process found within the Network Layer communicates with processes found at the other ends of all communication links connected to the transmitting node. To accomplish this, the Network Layer places information taken from the Transport Layer into the header portion of the data packet. With this, the header contains protocol information used by the peer Network Layer process.

From there, the Network Layer passes the packet to the Data Link Layer for placement into a frame. If the data travels to an intermediate node, the Network Layer in that node has the responsibility for forwarding the packet to the proper destination node. In some instances, the Network Layer must have the capability to handle packets to and from different types of nodes. In turn, this capability relies on the capability of the Network Layer to work with different communications protocols and types of addressing.

Theory to Function: The Network Layer

However, when we move to the Network Layer in both models, the differences become even more apparent. In figure 4.2, the Network Layer has a path to the LAN protocols and to the Internetwork Packet Exchange, or IPX layer. The IPX establishes the connectivity between networks and works as part of the network operating system. In comparison, the Network Layer shown for the AppleTalk model only has a path to the Datagram Delivery Protocol. With datagrams, the network does not establish a session before sending the data packet from one node to another.

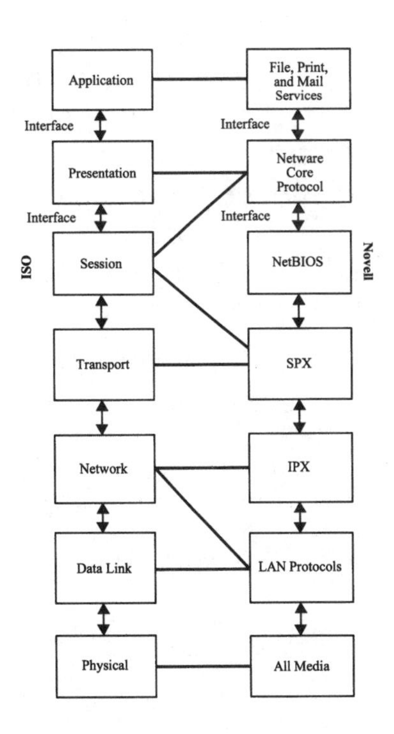

Figure 4.2. OSI model for Novell networks

With this model, AppleTalk supports communication between two sockets. In the most basic of terms, a socket functions as an interface for communication between a system and a network. A socket consists of a port and an address. During network operation, application programs use sockets to communicate across a network to a peer process.

Switching Between Layers

We can also categorize switches according to the OSI layer at which the device filters and forwards, or switches, frames. The switch categories include Layer 2, Layer 2 with Layer 3 features, and multilayer. A Layer 2 switch performs switching and filtering based on the OSI Data Link Layer MAC address. Along with those characteristics, the Layer 2 switch also functions similar to a multiport bridge but has a much higher capacity and supports full-duplex operation. As with bridges, a Layer 2 switch remains completely transparent to network protocols and user applications.

A Layer 2 LAN switch with Layer 3 features makes switching decisions based on more information than just the Layer 2 MAC address. Even though the switch includes Layer 2 switching functions, it also may incorporate Layer 3 traffic-control features. Some of those features include broadcast and multicast traffic management, security through access lists, and IP fragmentation.

A multilayer switch makes switching and filtering decisions on the basis of data link layer and network-layer addresses. During operation, this type of switch dynamically decides whether to switch or route incoming traffic. A multilayer LAN switch switches within a workgroup and routes between different workgroups.

Switched Networks and Microsegmentation

Most network administrators strive for microsegmentation — or the fewest possible users per network segment.

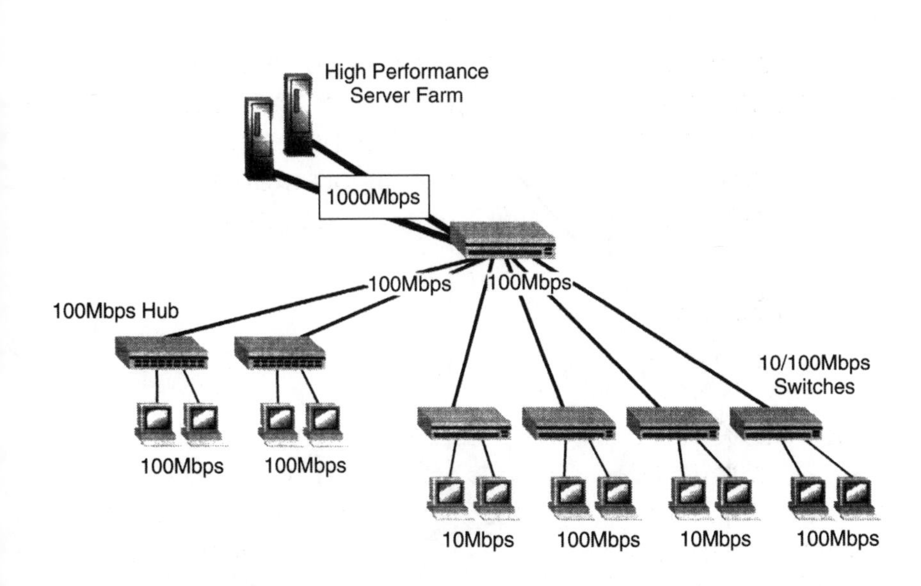

Figure 4.3. A LAN switch provides dedicated bandwidth to workstations.

Microsegmentation creates dedicated segments with one user per segment through the use of a LAN switch. With microsegmentation in place, each user receives instant access to the full bandwidth and does not have to contend for available bandwidth with other users. As a result, the collisions that normally occur in a shared Ethernet environment that depends on hubs do not occur.

During operation, the switch forwards frames based on either the Layer 2 address or Layer 3 address of the frame. Because of this, we can refer to a LAN switch as a frame switch. Ethernet, Token Ring, and FDDI networks take advantage of LAN switching. Figure 4.3 shows the configuration for a LAN switch that provides dedicated bandwidth to devices and illustrates the relationship of Layer 2 switching to the OSI Data Link Layer.

Routers

Although bridges connect two LANs for the purpose of forming a larger network, a bridge cannot handle the routing and session control functions needed in enterprise-wide networks. In

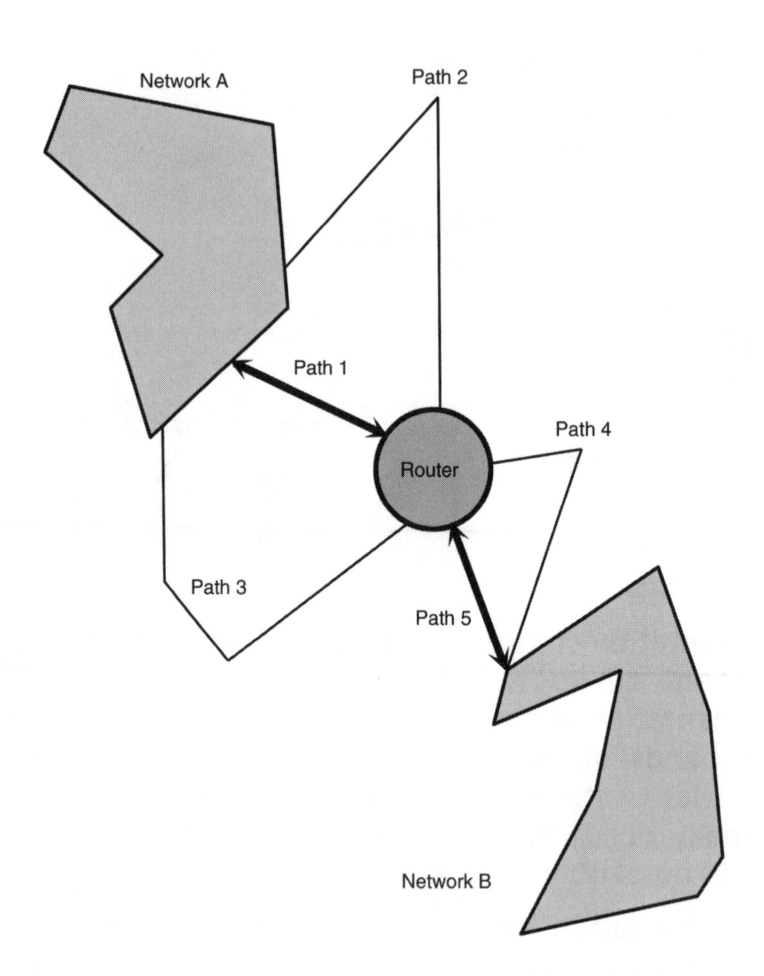

Figure 4.4. Networks connected by routers

contrast to bridges and as shown in figure 4.4, routers join local-area networks at the Network Layer and the Internet sublayer. A router converts LAN protocols into wide-area network protocols and performs the reverse process at a remote location.

Because routers offer more embedded intelligence than bridges, routers interpret more of the information found in a transmitted frame and support multiple protocols. The use of structured addressing allows routers to use redundant paths effectively and determine optimal routes even in a dynamically changing network. When network congestion or disruption occurs, the router can reroute traffic around the point of congestion or failure. Routers adapt quick to changes in network traffic by balancing the data load over the available routes.

Given these capabilities, many network administrators have replaced bridges with routers. Routers can identify the source and destination of transmitted data and build address tables to define the layout of the interconnected networks. The internetworking information becomes shared among various routing devices connected to the network. Each router on the network maintains a routing table and moves data along the network from one router to the next through the use of routing protocols.

The protocols include the Open Shortest Path First, or OSPF, protocol; the Intra-autonomous System to Intra-autonomous System, or IS-IS, protocol; the External Gateway Protocol, or EGP; the Border Gateway Protocol, or BGP; the Inter-Domain Policy Routing, or IDPR, protocol; and the Routing Information Protocol. Although the Routing Information Protocol had become a standard for many administrators, the expansion of routing capabilities has gone past the support offered through RIP. In contrast, OSPF includes an update procedure that requires each router on the network to transmit a packet with a description of the local links to all other routers. After receiving each packet, each router acknowledges the packet; OSPF builds distributed routing tables from the accumulated descriptions. When a link fails, the updated information floods the network and allows all routers to build new tables simultaneously.

Because routers use structured Layer 3 addresses, the devices can use techniques — such as address summarization — to build networks that maintain performance and responsiveness as growth occurs. In addition, routers offer significant advantages over bridges and switches. Because every data packet must go through the root bridge of a spanning tree, bridging and switching may result in nonoptimal routing of packets. While routers ensure network scalability, the devices also provide

- broadcast and multicast control,
- broadcast segmentation,
- security,
- quality of service, and
- multimedia.

Static Routing

Two types of routing exist. With static routing, a network manager configures the routing table to set fixed paths between two routers. The paths do not change after the action by the network manager. Although a static router recognizes that a link has shut down and has the capability to issue an alarm, it will not automatically reroute traffic.

Dynamic Routing

A dynamic router automatically reconfigures the routing table. To accomplish this, the dynamic router recalculates the lowest cost path with respect to load, line delay, or bandwidth. Some dynamic routers also rebalance the traffic load across multiple links.

Internet Protocol and the Network Layer

The Internet Protocol operates at the Network Layer of the OSI model and has rapidly become the standard for transporting data, voice, and video across networks. Every device connected through TCP/IP has a unique 32-bit address that identifies a network on the Internet and a host segment attached to a specific network. In addition, the Internet Protocol defines the packet for formatting. The packet format consists of a predefined data unit that has fields for addresses, checksums, control information, and data.

IP addresses separate into the network identifier and the node identifier. The network identifier designates the logical grouping for a user, and the node identifier gives each end user a unique identification on the network. Because the network identifier establishes common identities for IP devices, several devices can operate on the same physical network. The numeric IP addresses are classified as follows:

- Class A is an address structure supporting networks with more than 16 million nodes.
- Class B is an address structure supporting networks with up to 65,536 nodes.
- Class C is an address structure supporting networks with up to 256 nodes.
- Class D is an address structure supporting multicast address ranges.
- Class E is an address structure reserved for future use.

Logical addressing in the Internet Protocol defines the different classes of IP addresses according to the portion of the 32 bits used to establish the network identifier. IP addresses use the dotted decimal form to designate each portion of the 32-bit address. With the address separated by periods, each of the digits represents eight bits of the total. As figure 4.5 shows, each class type allocates a different number of bits to define a logical network identifier and uses the remaining number of bits for the nodes and subnets.

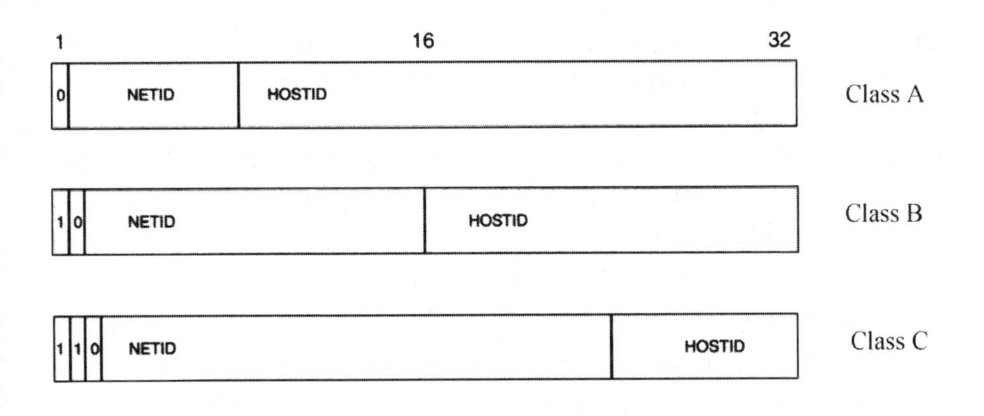

Figure 4.5. Illustration of IP class types

Network Masks

The Internet Protocol addressing scheme groups users together in networks with different sizes according to the class of the address. Each end user also receives a network mask that defines the number of bits used to define the network identifier. The network mask maps in binary form to define the bits used to describe the network and becomes part of the route determination process.

Subnets

The masks for the Type A, B, and C classes also support subnetting. When a network requires further segmentation of a single class A, B, or C major network, subnetting provides the mechanism needed to use a large, single address space to describe many smaller separate areas. As an example, the subnetting of a large Class A network can establish 255 Class B networks or 64,000 Class C networks. With a natural mask of 255.0.0.0, a user becomes described as one member within a group of 16 million addresses. With an 8-bit subnet mask, the same user becomes a member within a logical group of over 64,000 addresses.

Secondary IP Addresses

Secondary IP Addressing becomes useful for constructing large switched networks that contain more nodes than handled by one subnet or network. The process of secondary addressing adds more IP addresses and networks to a single router interface. Secondary IP Addressing can free routers from forwarding responsibilities and allow the use of routing resources to handle the proper routing into and out of the switched area.

Address Resolution Protocol and the Network Layer

Different protocol suites may use either the Address Resolution Protocol, or ARP; the Hello protocol; or the embedded Network Layer MAC addresses for determining the MAC address of a device. The Address Resolution Protocol maps network addresses to MAC addresses. This process occurs through the implementation of the Address Resolution Protocol by many protocol suites. When a network address becomes associated with a MAC address, the network device stores the address information in the ARP cache. In turn, the ARP cache enables devices to send traffic to a destination without creating ARP traffic.

Depending on the network environment, the process of address resolution may feature slight differences. As an example, address resolution on a single LAN begins when End System A broadcasts an ARP request onto the LAN in an attempt to learn the MAC address of End System B. Although End System B replies to the ARP request only by sending an ARP reply containing its MAC address to End System A, all devices connected to the LAN receive and process the message. End System A receives the reply and saves the MAC network addresses. Whenever End System A must communicate with End System B, the device checks the ARP cache, finds the MAC address of System B, and sends the frame directly without first having to use an ARP request.

This process changes with the interconnection of the source and destination devices found on different LANs through a router. End System Y broadcasts an ARP request onto the LAN in an attempt to learn the MAC address of End System Z. Again, all devices on the LAN — including the router — receive and process the broadcast. The router acts as a proxy for End System Z by checking its routing table and determining that End System Z operates on a different LAN.

In response, the router replies to the ARP request from End System Y and sends an ARP reply containing the router address as if the address belongs to End System Z. From there, End System Y receives the ARP reply and saves the MAC address of the router in its ARP cache within the entry for End System Z. When End System Y must communicate with End System Z, the device checks the ARP cache, finds the MAC address of router, and sends the frame directly without using ARP requests. The router receives the traffic from End System Y and forwards it to End System Z on the other LAN.

Hello Protocol and the Network Layer

Another Network Layer protocol called the Hello protocol enables network devices to identify one another and indicate functional status. During operation with the Hello protocol, a new end system powers up and broadcasts Hello messages onto the network. Devices on the network then return Hello replies. In addition, the devices send Hello messages at specific intervals to indicate the functional status.

Network devices can learn the MAC addresses of other devices by examining Hello protocol packets. The Xerox Network Systems, or XNS; Novell Internetwork Packet Exchange, or IPX; and DECnet Phase IV protocol suites use predictable MAC addresses. The MAC addresses are predictable because the network layer either embeds the MAC address in the Network Layer address or uses an algorithm to determine the MAC address.

The Transport Layer

The Transport Layer resides as the fourth layer in the OSI model. In brief, the Transport Layer sees the entire network and uses the Physical, Data Link, and Network Layers to establish end-to-end communications for the higher levels of the model. End-to-end communications involve peer processes at either end of the connection communicating through a common protocol. As a result, the Transport Layer has the primary task of moving messages from one end of a network to the other end. Processes within the Transport Layer recognize all intermediate nodes as adjacent nodes and rely on the lower levels of the OSI model to pass the data through the intermediate nodes.

Network Layer to Transport Layer Services

As we discovered in the opening chapter, we can classify networks according to the area that the network covers. As an example, a wide-area network may cover a campus, a city, or an entire region. Networks may consist of subnetworks that link together to form the larger networks. Since the Network Layer exists at the junction between subnetworks and full networks, it provides a number of services for the Transport Layer.

During the interaction between the Network and Transport Layers, the Network Layer establishes a unified addressing scheme. As a result, each node on the network has a unique address that becomes part of a total, consistent addressing scheme for the network. In addition, the integration of Network and Transport functions also facilitates the use of the circuit-switched and packet-switched networks, which we will discuss in later chapters. While a circuit-switched network establishes virtual circuits between pairs of nodes and transfers data in both directions across the circuits, a packet-switched network relies on the sending of packets that contain the full network address of the destination node.

Transport Layer Services

Outside of the interaction with the Network Layer, the Transport Layer also provides a set of network services that include addressing, managing connections, controlling data flow, and buffering. During operation, several processes may occur simultaneously within a node. With addressing, the Transport Layer ensures that a connection occurs to each of the specific processes. Connection management establishes and releases the connections.

Data flowing on the network must travel at a rate accepted by the network hardware. With flow control, the Transport Layer considers the capacity of the hardware to handle data transmission and adjusts accordingly. In relation to flow control, the Transport Layer also ensures that the correct amount of memory remains available for buffers in the receiving node.

The Transport Layer also matches the resources found in the lower layers to the upper layers of the OSI model. With some applications, the upper layers may require slower services that those provided by the channel. To accommodate this requirement, the Transport Layer uses multiplexing, or the interleaving of different message packets. At the receiving end, the Transport Layer sorts the packets and recreates the original message by using user IDs stored in the message headers.

Other applications may establish a condition where the upper layers require faster service than that available with a single channel. Paralleling the flow of data across the channel multiplies the effective flow of the data for the higher layers. Generally, an operating system such as UNIX will provide the multiprogramming capability that allows the lower layers to operate in parallel.

Transmission Control Protocol and the Transport Layer

The Transport Layer allows peer entities located on the source and destination hosts to communicate. Using the connection-oriented Transmission Control Protocol, or TCP, the Transport Layer delivers the data stream originating on one computer to any other machine connected to the Internet without error. The sending TCP process breaks the data stream into discrete messages and passes each message onto the Internet Layer, whereas the receiving TCP process reassembles the received messages into an output stream. Along with TCP, the Transport Layer also uses the connectionless User Datagram Protocol, or UDP, for applications that do not rely on the Transmission Control Protocol.

The Session Layer

After the Transport Layer establishes a connection between two nodes, the Session Layer sets a number of parameters that allow actual communication between the nodes to occur. The parameters apply not only to the nodes but also to the higher layers of the OSI model. Before the transfer of information takes place, the Session Layer ensures that communication can occur and that the nodes do not attempt to communicate simultaneously. Then, the Session Layer manages the communication by breaking it into usable parts. When the communication ends, the Session Layer provides an orderly method for the break to occur.

During operation, the Application Layer may need to access the services given by the Session Layer. When the Session Layer provides services for the Application Layer, the services pass through the Presentation Layer as part of a suite. Regardless of the need for one or all services, the suite passes through as a whole.

Sessions

We define the session as simply the communication that occurs between two nodes after a connection has taken place. Several sessions may occur during one connection or one session may require the establishment of several connections. The Transport Layer allows the multiplexing of several sessions onto a single connection. Sessions occur through dialogs that begin the communication and provide an orderly ending to the communication.

Simultaneous communication in both directions occurs. However, the communication remains orderly through the use of tokens. As the communication takes place, the nodes pass the token back and forth. The node with the token can transmit data while the node waiting for the token receives the data. With this configuration, the passing of the token signals the nodes that a change in the direction of data flow will occur.

The Session Layer also manages the starting and stopping of the data transmission. During the transfer of information from one node to another, the receiving node must obtain notification that the transmission has reached the end of the file. Control information embedded in the file header informs the node that the transmission has reached the end of the file. Without the activity management given by the Session Layer, several different files transferred during one session could appear as one large file. In this sense, each file becomes an activity.

The Presentation Layer

Computer systems interpret data according to different character codes. As an example, an IBM mainframe communicates with a different set of codes than an Apple Macintosh computer. As the data flows from one type of system to another, the Presentation Layer translates the different character codes so that the systems may communicate.

Character coding involves the placement of command sets within a structured code. In addition, the format of the document, the format of floating point numbers, or the type of program used to compile the language data transmitted from computer to computer may vary. When considering the format of the document, a special language format may display a document in a certain manner. The format of floating point numbers consists of a given number of bits in a specific order within a word or double word. When moving from computer to computer, the number of bits varies. During the compiling of data, the representation of data may have different forms.

In addition, the Presentation Layer encrypts, decrypts, and authenticates the data to prevent unauthorized access to information and to confirm the source of the information. The Presentation Layer also compresses data passed by the Application Layer to save space within the channel during transmission. At the receiving end, the Presentation Layer decompresses the data.

The Application Layer

At the highest level of the OSI model, the Application Layer contains network applications. As we have worked through the remainder of the OSI model, we found that each layer offered a variety of services. Network Applications that include electronic mail, file transfer, and file sharing software use the services to accomplish specific tasks. Because each application relies on protocols for communication, the Application Layer must have a method for accessing services for the protocols. To do this, the Application Layer passes unmodified messages from the application to the Presentation Layer.

Utilizing the Transport, Session, Presentation, and Application Layers

The same comparison becomes effective when viewing the Transport, Session, and Presentation Layers for both models. Looking back at figure 4.2, note that the Transport Layer has a path to the Sequential Packet Exchange (SPX) Layer while the Transport Layer for the AppleTalk model has a path to the AppleTalk Transaction Protocol Layer. Within the Novell terminology, the SPX Layer provides the Transport Layer functions for a Novell NetWare System.

The AppleTalk Transaction Protocol, or ATP, conducts end-to-end interaction within an AppleTalk network and provides request-response transactions. ATP interactions involve independent transactions on the network. In addition, the transactions found within the ATP layer establish the basis for services seen in the upper layers of the AppleTalk model.

Again moving back to figure 4.2, the Session Layer for the Novell OSI model has a path to both the SPX and the NetWare Core Protocol. The Session Layer shown for the AppleTalk model has paths to the AppleTalk Session Protocol layer and the AppleTalk Filing Protocol layer. Novell's NetWare Core Protocol transmits information between clients and servers. Going back to the IPX Layer of the Novell OSI model, the IPX Layer provides Network Layer functions by transporting the NCP messages. The AppleTalk Filing Protocol Layer allows workstation users on the AppleTalk network to share files.

The Presentation Layer for the Novell OSI model connects only to the NetWare Core Protocol, but the Presentation Layer for the AppleTalk model has paths to the AppleTalk Filing Protocol and the File, Print, and Mail Services. Both models have the Application Layer connected to the File, Print, and Mail Services. Application programs request and manipulate files by using native file system commands found within the operating system of the workstation. Printers connected through a network allow users to share print resources, but print sharing applications ensure that no bottlenecks occur within the process.

In contrast to the OSI models shown for Novell NetWare and AppleTalk networks, the relationship between Microsoft Windows NT and the model remains relatively basic. The network operating system provides transparent support for client applications in a distributed environment. Moving to figure 4.6, Windows NT supports network interface cards and connectors through the Physical Layer and network protocols such as Ethernet and Token Ring through the Data Link Layer.

Low-level software applications called device drivers link devices used at the Physical Layer to the protocols found in

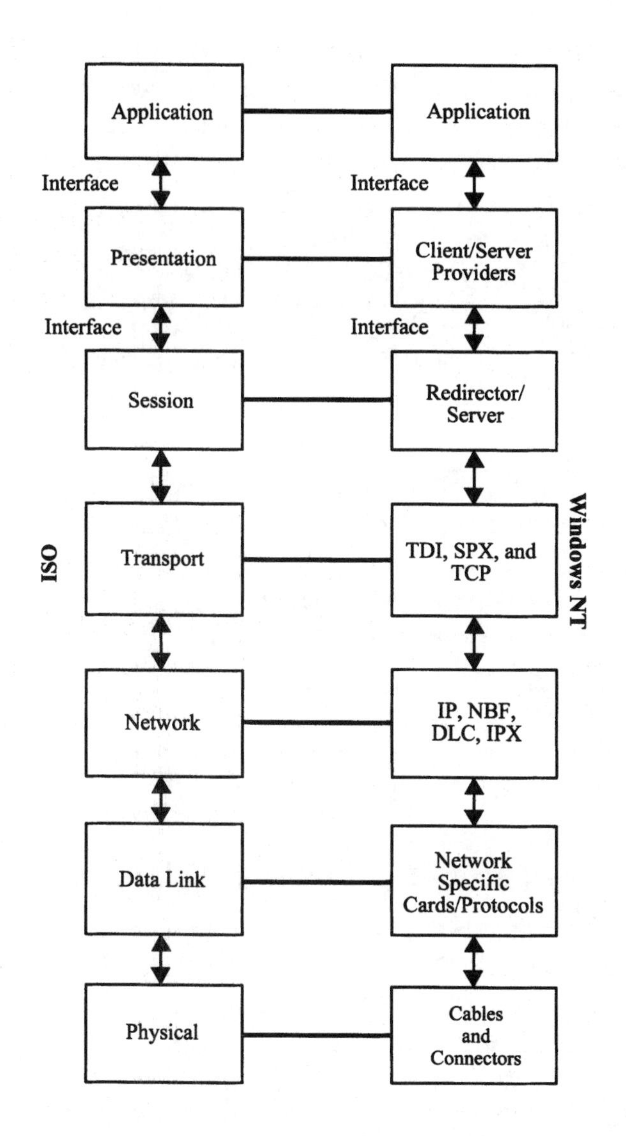

Figure 4.6. OSI model for Windows

the Network Layer. Microsoft Windows NT uses the Network Device Interface Specification to provide the linkage. As seen with the Novell OSI model, the Transport Layer supports SPX. In addition, the Transport Layer also supports the Transmission Control Protocol and the Transport Device Interface, or TDI. The TDI allows the software to establish an interface with multiple transport protocols.

At the top of the Microsoft OSI Model, the Session Layer allows the loading of redirectors for Windows NT, Novell NetWare, and Banyan VINES. Redirectors work in a client-server environment and decide whether a request for some type of computer service should involve an individual work-station or the network server. The Windows NT application layer supports client-server applications as well as peer-to-peer processing, while the Network File System, or NFS, protocol allows servers to share disk space and files.

Simple Mail Transport Protocol and the Session, Presentation, and Application Layers

Internet-based electronic mail connections occur through a source system establishing a TCP connection at a destination system. The Simple Mail Transport Protocol, or SMTP, accepts the incoming transmission and copies the electronic mail messages into the appropriate destination mailboxes. An SMTP server transfers text-only electronic mail between host servers along with a line of text that provides an identity and a ready-to-receive confirmation.

File Transfer Protocol and the Session, Presentation, and Application Layers

The Internet relies on the File Transfer Protocol, or FTP, to access and transfer files from a host server to another host or to a client. FTP servers located on global networks allow users to log in and download files. Because the FTP interface remains rather difficult to use, many Web-based applications have begun to shift to HyperText Transfer Protocol, or HTTP, servers for file transfers.

Gateways and the Application Layer

Bridges and switches cannot interconnect nodes that use different architectures. In comparison, a gateway allows communication between different architectures and application level protocols as shown in figure 4.7. In addition, a gateway per-

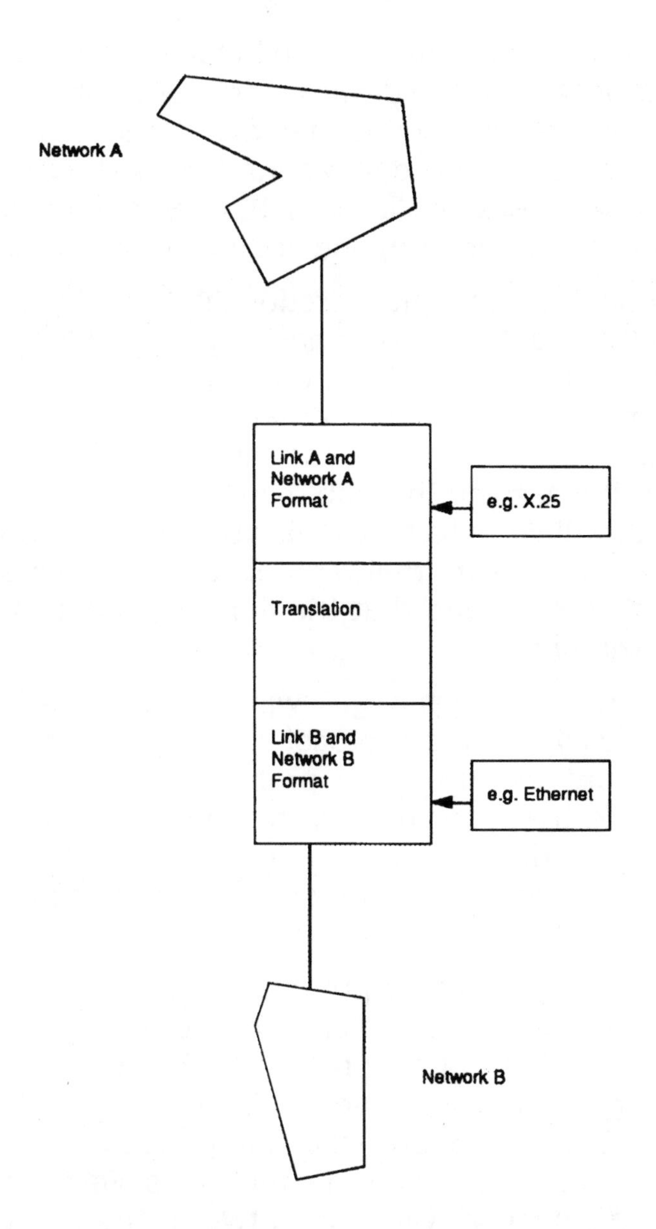

Figure 4.7. Networks connected by gateways

forms protocol conversion through all layers of the OSI model. Because a gateway provides all the functions given by routers, the term "gateway" often references the use of a router. However, gateways primarily operate as LAN-WAN connections, perform more sophisticated functions than routers, and have a more difficult installation process.

When used as a LAN-WAN interface, a gateway resides on both the LAN and wide-area network. At the LAN, the gateway appears as a server device. At the WAN, the gateway appears as a host-dependent terminal or workstation controller. Gateways facilitate LAN workstation connections to a variety of host, private, and public network environments.

Although the additional functionality may result in slower network throughput, gateways not only connect different types of networks but also ensure the compatibility of data transported from one network to another. In addition to translating data, a gateway can also extract specific information about the data traffic passing through the gateway and evaluate the status of data links interfaced with the gateway. As a result, the gateway can ensure that the network links remain reliable while handling data and that the links maintain user-defined error rate thresholds.

Many different types of gateways exist. For example, one type of gateway called a protocol converter changes the protocol of one architecture to the protocol of another. To accomplish this task, the protocol converter operates above the Data Link Layer. In addition, the operation of the protocol converter must remain transparent to any processes found in the layers located at each end of the connection.

Because gateways have protocol-specific functions, a gateway can provide access to a mainframe computer. Multiprotocol gateways in the form of adapter cards link stations on a local-area network to a mainframe through an SDLC link and an X.25 connection to a packet-switching network and allow the stations to function as mainframe terminals. Following the connection to the packet-switching network, the gateway either converts the LAN traffic or routs the traffic to a packet network node and transmits from that node to the destination in the X.25 format.

An electronic mail gateway converts documents from one e-mail format to another e-mail format. If a business wishes to attach an internal electronic mail network to the Internet, a gateway converts the internal mail documents for transportation through the Simple Mail Transport Protocol found within TCP/IP. In turn, the gateway also converts incoming mail delivered from the Internet into the format used on the internal electronic mail network.

Packet-Switched Networks and the X.25 Format

With packet switching, a data packet travels from a source node of a network and through intermediate nodes before arriving at the destination node. Packet switching provides a method for network devices to share a single point-to-point link to transport packets from a source to a destination across a carrier network. Ethernet networks, Asynchronous Transfer Mode networks, Frame Relay, Switched Multimegabit Data Service, and X.25 are examples of packet-switched WAN technologies.

When the packet travels through the intermediate node, or packet switch, the node switches the packet to the next node in the sequence. We can refer to packet-switched networks as connectionless networks because no physical connection occurs between the source and destination nodes. Packet-switched networks that use the X.25 protocol exist on the leased switching. An X.25 network includes the following:

- the physical hardware and cabling used to make a physical circuit that connects devices together,
- a complete path between two communicating devices used to make a virtual circuit; and
- a logical connection between the user node and the network.

5

Servers

Introduction

Server technologies have evolved from the traditional client-server networks to encompass Web delivery, proxy access, printing, facsimile delivery, communications, and remote access. This chapter provides a tour of the server technologies and explains the processes used within the servers. It explains the differences between processors used within the servers and defines different server applications. In addition, it covers fault tolerance as well as server load balancing. The chapter concludes with a discussion about server clustering.

Defining Servers

A computer operating as a server works on a network and facilitates resource sharing. Server computers usually contain larger amounts of random-access memory and higher capacity disk drives than standard computers on the network. As a result, the server supports concurrent processes and heavy processing tasks. In brief, a server distributes files, messages, and applications to workgroups and stand-alone computers in remote locations. The use of a server and the distribution of resources through the server minimize the possibility of a failure in centralized resources.

A server may operate as a dedicated or shared function microcomputer, a high-powered workstation, a minicomputer, or a mainframe. The network operating system controls the functions and operation of the server. Protocols combine with the operating system to establish a framework for communication across the network.

When preparing to purchase a server, consider the number of processors, the bus architecture, the main memory, the amount of memory cache, the type of storage devices, the amount of storage provided through the devices, the number of bays, and the number of I/O ports. Later we provide additional information about processors typically used in servers. Even though the bus architecture includes the Industry Standard Architecture (ISA), Peripheral Component Interconnect (PCI), or Extended Industry Standard Architecture (EISA), the main memory refers to the amount of random-access memory contained within the computer and the memory cache refers to the amount of required wait time for program initialization.

Servers may include a variety of storage devices such as disk drives, tape drives, and optical disk drives. Extra bays in the server allow the installation of additional disk drives or other peripherals. Most servers offer input/output expansion slots that accommodate the attachment of devices such as printers and scanners.

Processors

Microprocessors differ in the amount of instructions executed, the number of bits processed within a single instruction, and the number of instructions that the processor can execute per second. Microprocessors employ either the complex instruction set computing, or CISC, architecture or the reduced instruction set computing, or RISC, architecture. Intel x86 and Pentium processors are prime examples of CISC technology. The limited command set and simpler architecture of RISC processors allows those devices to execute instructions three or more times faster than a CISC processor at any given clock speed.

CISC-Based Microprocessors

Although RISC-based microprocessors have a traditional place in the embedded systems market, manufacturers of CISC-based processors have begun to reshape their products for the embedded market. According to industry marketing managers, CISC-based processors provide a stable, known architecture for embedded development and the capability to allow developers to bring products to market quicker. Single-board, PC-compatible computers can implement market-attractive features while accelerating software development. Designers can shorten the time to market by taking advantage of software resources currently available for the mainstream processors.

In addition, the CISC-based processors such as the Intel Pentium family offer greater performance and higher processing speeds. As an example, Intel has enlisted a group of third-party developers in cooperative support activities such as tools, operating systems, and BIOS. CISC-based processors such as the 80C186, the Intel386, the Intel486, and the Pentium family of microprocessors have been used in embedded applications such as Internet appliances. During late 1997, Intel announced the formation of embedded applications support for existing 200-MHz Pentium MMX processors. The new embedded application Pentiums feature double 32 kilobytes of on-chip cache, enhanced branch prediction and pipeline, and deeper write buffers.

RISC-Based Microprocessors

Because new applications continue to evolve, the demand for high-speed RISC microprocessors has grown substantially. The performance gains seen within embedded microprocessor technologies — where the number of instructions per second has grown from several million to over 2 billion instructions per second — have occurred because of the capability to place millions of transistors on a single chip. In turn, this advance has led to superscalar architectures with multiple-integer and floating point execution units, as well as high levels of pipelining.

Superscalar architectures that have multiple execution units provide the capability to execute two or more instructions simultaneously. Superior pipelining schemes add pipeline stages to the processor and allow the placement of multiple instructions in the CPU execution queue. As a result, new instructions in the pipeline begin with each clock cycle.

Each of these architectural improvements has led to the addition of more complex instruction sets to superscalar processors. Because of the increase to 128-bit wide buses in microprocessors such as the Alpha, CPU word lengths have grown from 16 to 64 bits. With this, 64-bit data words can be divided into two, four, and eight subwords. As a result, a processor involved with digital signal processing, image processing, or multimedia applications can perform as many as eight parallel computations on the data words. The improved computational power also results from minimum clock speeds of 200 MHz to maximum clock speeds of 600 MHz and higher along with exceptional integer and floating point performance. At the high end, data throughput may reach 2 billion instructions per second. Each of the improvements in computational power and throughput also requires the addition of large caches, the use of dynamic execution control, the implementation of multiple execution units, and advances in register logic and branch prediction.

Fault Tolerance in Servers

Network servers include features that guard against the loss of data. Fault tolerance in terms of hardware becomes defined as maintaining 100 percent availability, safeguarding critical data, and maintaining user productivity during the failure of any network component. Network administrators rely on hardware and software fault tolerance solutions.

Hardware Fault Tolerance

The hot standby solution to fault tolerance duplicates the CPU, ports, network interfaces, memory expansion cards, disks, and input/output channels. As a result, an alternate hardware component can assume responsibility if a network failure occurs. The secondary system monitors the tasks of the primary system and duplicates the tasks. In contrast to the hot standby system, the use of a load-balancing mechanism allows all hardware components to function simultaneously. The load-balancing mechanism reallocates processing tasks to other components if a failure occurs. However, the load-balancing system depends on a more sophisticated operating system that can continually monitor the system for errors. When a component failure occurs, the operating system adapts to the problem by dynamically reconfiguring the system.

Software Fault Tolerance

Fault tolerant software works in conjunction with server hardware to ensure the availability of data and to maintain server operation during a failure. During operation, the fault tolerant software switches processing tasks from primary to secondary hardware components. With the hot standby solution, the software shifts complete control to the standby system after detecting an error. The switch occurs regardless of the type of component failure. With the load-balancing system, only the component experiencing the failure becomes replaced by the secondary component.

Near Fault Tolerance

A near fault tolerant system features servers that monitor one another through a standard RS-232 connection or a network link. Each system maintains a log of error messages generated by the operating system used by the other unit. Moreover, both servers share the same disk drives. When a catastrophic error occurs on one server, the other server assumes control and accesses the applications and files held on the primary server.

Server Technologies

Mainframes

A mainframe computer functioning as a file server can manage large amounts of data generated by network users and network management software. In addition, a mainframe can store data in a data warehouse, distribute processing tasks to connected workstations, collect results, and consolidate files. However, the use of a mainframe as a server shifts a portion of the processing responsibilities back to the individual workstations.

Superservers

Originally developed during 1989, a superserver takes advantage of multiprocessor designs. Multiple processors may operate with a single bus for the entire system or use multiple buses that eliminate contention. Computers configured for use with multiprocessor technologies usually include fast-wide or ultrawide SCSI RAID controllers, large amounts of processor cache memory, large amounts of error-correcting RAM, Fast Ethernet cards, and reporting systems. Along with supporting large data transfers, superservers also support the implementation of all network operating systems and protocols.

In addition, superservers include an input/output structure similar to that seen with mainframe computers but they have a higher packet throughput. Because the I/O processors implement control logic and concentrate data, the central system processor does not contend with single-byte data accesses, LAN protocols, or the location of data on the disk drive. Moreover, the configuration of the superserver hardware allows system administrators to centralize and compartmentalize LAN administration on one server.

Single-Processor Servers

In most scenarios, a standard single-processor microcomputer can perform server functions. Much of the performance has become possible because of improvements in processors, bus designs, and disk technologies. Microcomputer-based servers also benefit from the low-cost availability of higher speed processors. Although microcomputer-based servers lack the capabilities of mainframes or superservers, microcomputers offer the flexibility to accommodate any specialized need such as imaging, gateway connection, and facsimile. Figure 5.1 shows a microprocessor-based server.

Figure 5.1. A micro-processor-based server

Dedicated Servers

With a dedicated server arrangement, the network operating system runs on a dedicated computer and provides specific tasks for the network such as file sharing, communication, and administration. Because the server centralizes information, a network administrator sets user access policies, defines security rules and measures, establishes data protection, and maintains the operation of the network. Dedicated servers also run multiuser applications that allow many users to access resources and information at a central location. Dedicated server applications include databases, communication services, and electronic mail.

File Servers

A file server operates as a platform for storing data, applications programs, and files while allowing shared access to the programs or information through network connectivity. All data files contained within the server hard disk drive include access attributes set to read only or read and write. In addition, the server process allows the sharing of network files. A user may transfer an entire file from the file server to a workstation. The following attributes are associated with file servers:

- exclusive (the file cannot be shared),
- write access denied (the file can only be read by others), and
- deny none access (the file can be shared, read, and written to).

In most instances, the application assigns access rights but allows modification of the rights. Even though access control exists as the first step, the network operating system on the file server also controls the synchronization of the files. With this capability, file updates made by different users do not occur at the same time through the use of file or record locking.

File locking prevents multiple user updates in the form of writing to the file. The synchronization tool allows the sharing of the file but only for the purposes of reading. Record locking uses either physical or logical locks for file protection. A physical lock protects a record or a series of records within a file by preventing access to the records on the hard disk. A logical lock assigns a lock name to the protected records and allows access only after an application checks the lock name, the record attributes. If the application lacks the proper authorization, the logical lock prevents access to the record.

In addition to file and record locking, network operating systems also use semaphores to organize contention to disks, files, and records shared on the file server. The semaphore provides an intelligent bit flag that can accept a name, set, clearing, and testing. Compared to file or record locks, semaphores provide more flexibility. The flags apply to files, records, record groups, or shared peripherals. A semaphore may have any assigned meaning and can operate with different applications.

Database Servers

Database servers divide database processing into a front-end application that runs at the desktop and a back-end processing that runs in the database engine held on the server. During operation, the downloading and uploading of individual records eases locked data and network traffic. Retention of data at the server also facilitates backup and security operations for the database records.

Although a file server offers ease of use through its capability to share applications or files, limitations exist. As an example, an operation that requires access to a database may result in the transfer of thousands of records from the file server to the workstation. Transferring the large number of records over the network can cause a dramatic decrease in network performance.

Database servers solve the problem of passing an entire file over the network by processing the user request on the server and then transferring the results of the operation to the individual workstation. With this, the processing speed and storage capability of the server increase the performance of the network and decrease the time needed to answer the request. Using a database server places greater emphasis on the capabilities of the server rather than the capabilities of the network.

Because processing occurs on the server, one operation may affect the abilities of other users to complete an operation. The server must have the available processing power, memory capacity, and storage capacity needed not only to complete a requested operation efficiently but also to also maintain the operations of other users. In addition, an interface must exist between the server and the application software used to make the request. The interface translates the data needs of the application into the language used by the database server.

Most database servers also use transaction tracking or the temporary storing of a transaction in a buffer until the completion of an operation. With this type of rollback recovery in place, an incomplete transaction cannot corrupt a database. In addition to rollback recovery, some systems may use roll-forward recovery and complete the transaction through the use of an audit trail if an error occurs. Most database servers rely on hierarchical storage or the placement of often-accessed data on-line and archived data off-line.

Print and Communications Servers

Print servers use third-party software to control access to network printers. Shown in figure 5.2, print servers can affect the performance of the network because the server requires microprocessor cycles from a workstation or file server during

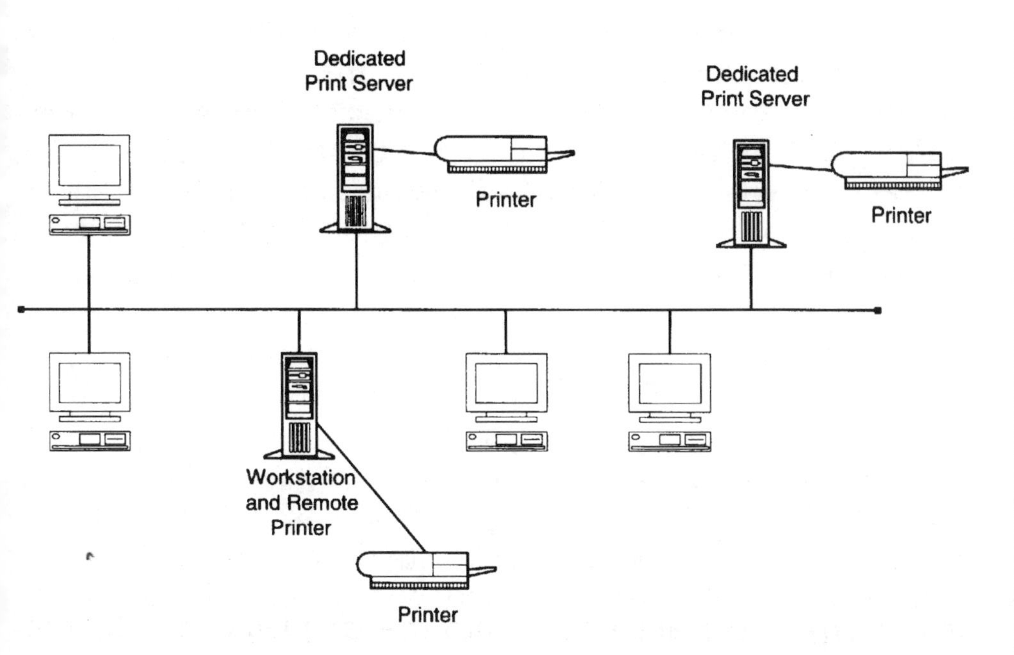

Figure 5.2. Diagram of a print server

operation. As with print servers, communications servers — also referred to as Remote Access Servers — allow the pooling of resources but operate with modems and phone lines. Supporting communication resources requires the installation of multiple serial ports and high interrupt rates at the server.

Remote Access Servers

Because of the need to connect network users from any location, remote access technologies have emerged as one of the newer aspects of local-area network strategies. Remote access increases the reach of the enterprise LAN and allows remote users to have the same access to network resources as local users. The typical users of remote access may telecommute from the same location and require connection to a corporate backbone on a regular basis. Because high costs accompany a dedicated connection, most telecommuters rely on some type of high-speed modem.

Other users have a mobile lifestyle and access the network through portable computers that have internal modems. Generally, a mobile user requires occasional access to a corporate network while working from a temporary site. Because of the nature of their work and the lack of a fixed workplace location, mobile users rely on analog dial-in access through public telephone networks.

Remote Access Technologies

Remote access technologies fall into two categories called the remote control and remote node. With the remote control technologies, a remote user dials into a dedicated host computer connected to the corporate backbone and assumes control of that computer. Remote control allows the remote user to run applications, access files, and transfer files that reside on the host computer. All applications remain on the host computer.

Because the host computer provides all processing, the transmissions between the user and host occur at a faster rate. The transmission speed also occurs at a faster rate because of the content of the transmission. Although the user inputs data, the only data transferred over the network link consists of screen updates, keystrokes, and mouse movements.

Remote control offers the advantage of granting access to applications located on the network, but it requires a computer for each user. In addition, the access for the user remains limited to applications that the host computer can access. Given the functions of remote control, the technology operates best with text-based applications that require fewer screen updates.

In comparison to remote control, remote node access includes a server component and the client component. The server component supports the network transfer protocol, and the client component consists of software running at the

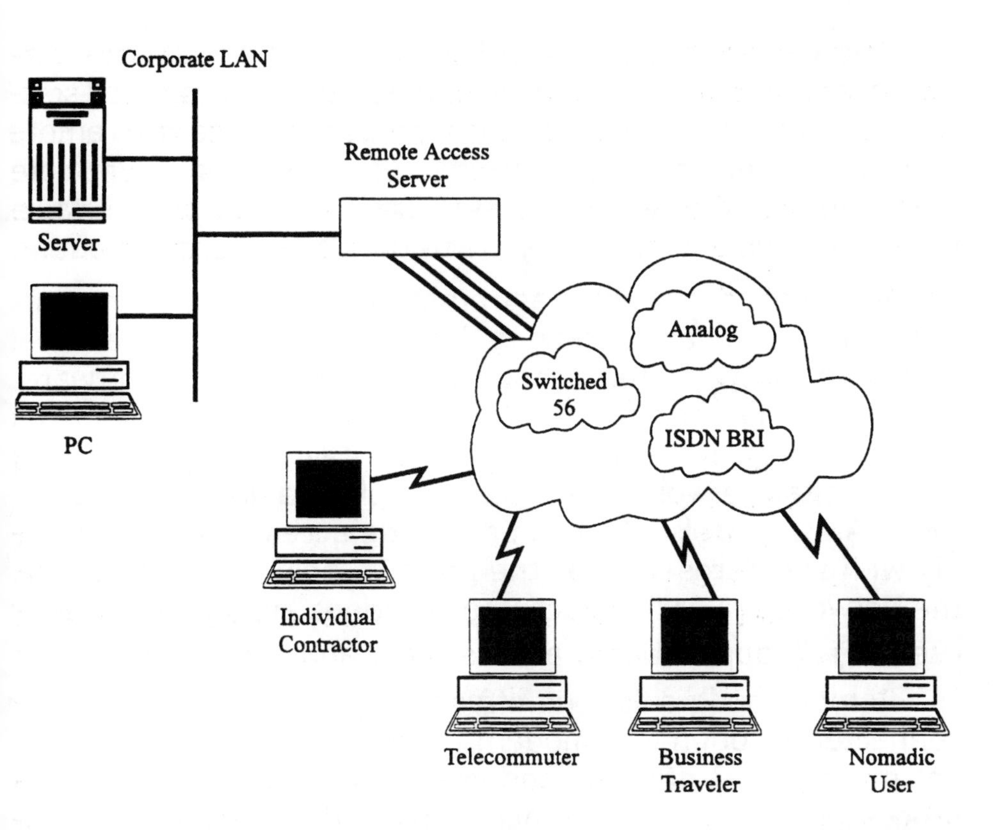

Figure 5.3. Diagram of a remote-access server

remote client computer. Called a remote access server, the server component contains router and modem technologies designed to transmit packets of data between systems as if the systems reside on a network. As pictured in figure 5.3, the remote server connects directly to the corporate network and monitors the modem pool, access security functions, and management functions.

Remote access servers can support both client-to-LAN and LAN-to-LAN connectivity. In addition, most remote access servers support the Ethernet and token ring standards. Most remote access server solutions support the Windows, Macintosh, UNIX, and OS/2 operating systems.

Remote node access allows the remote user to appear as a logical local link to the corporate network even though the actual physical connection occurs through a dial-in link. The remote access client software substitutes for the network interface card and driver. As a result, the remote access client software causes the application software to recognize the remote client as a direct connection to the network. When the remote user dials into the remote access server through an interface, the server bridges traffic directly to the corporate backbone.

Systems that operate as remote nodes range from communication software packages that require the use of a serial port to complete systems that include hardware and software. With this range of system in mind, functionality ranges from dial-out modem sharing, support for ISDN (Integrated Services Digital Network), remote control capabilities, and common LAN protocols. Most remote node access solutions comply with standards such as the point-to-point protocol, SNMP, the Network Driver Interface Specifications, and Open Datalink Interface interfaces.

Remote node solutions offer the best functionality when connected through fast modem technologies. The remote client must have sufficient random-access memory and microprocessor power to run applications and maintain the background modem connection. Because remote access depends on the capability of the system to move data through serial ports with little processor involvement, the remote access hardware must have the optimal processing power and serial port throughput.

Dial-up Protocols

The Point-to-Point Protocol, or PPP, has become the industry standard for transporting TCP/IP, AppleTalk, and IPX traffic from a dial-in device to a remote access server. PPP offers speed, flexibility, and security options while simultaneously supporting multiple network layer protocols over synchronous and asynchronous lines. In addition, PPP features link quality monitoring, or LQM, along with support for the Challenge-Handshake Authentication Protocol and the Password Authorization Protocol.

The Serial Line Interface Protocol preceded the introduction of PPP and operates as asynchronous protocol for running TCP/IP over serial lines. Just as the Apple Remote Access Protocol allows remote Macintosh users to dial into an AppleTalk LAN, the Multilink Point-to-Point Protocol establishes a virtual path between sites. The Multilink Point-to-Point Protocol aggregates telephone lines or ISDN B channels and facilitates the transmission of large amounts of data.

Drafted by eight leading remote access vendors, the Bandwidth Allocation Control Protocol, or BACP, aggregates channels and uses on-and-off bandwidth capabilities for the purpose of saving users time. BACP permits devices to combine the two 64-kbps B channels on an ISDN line to form a 128-kbps connection. As bandwidth needs change, BACP adds or drops dial-up and ISDN lines. Multilink PPP requires the user first to dial up each channel manually, but BACP dynamically adds and subtracts the channels.

The Point-to-Point Tunneling Protocol, or PPTP, establishes virtual private networks by allowing users to access corporate networks through the Internet. Rather than requiring the dialing of a long-distance number to access the corporate network, PPTP allows a user to dial a local number into an Internet Service Provider (ISP). Data packets wrap into a tunnel through PPTP and travel through the Internet into the remote access server. The remote access server routs, decrypts, validates, and filters the packets into the corporate network.

Serial Communication

Serial communications transmit data one bit at a time and occur in either a synchronous or asynchronous format. Synchronous communications involve the use of clock pulses to synchronize the transfer of data. Asynchronous communication transmits data intermittently rather than in a steady stream. Both synchronous and asynchronous communications rely on sync bits that acquire and maintain the synchronization between the sending and receiving devices.

Synchronous Communication

With synchronous communications, a constant flow of data allows each piece of data to remain ready for a data transmission. Each character represents either actual data or an idle character. Because synchronous communications do not mark the beginning and end of each data byte, faster data transfer rates occur. Before modems begin to transmit and receive data, the modems use a set of synchronized clock pulses to set the internal timing of clock circuits.

At the beginning of the established connection, each modem transmits a burst of bits that has a specific length. During the sending of the data, the transmitting modem places a one or zero on the line at established intervals. The receiving modem samples the line on the same timetable and transmits the condition of the line to the other modem. The modems must remain synchronized so that communication can occur.

Asynchronous Communication

Compared to synchronous communications, asynchronous communication does rely on an established timetable for the transmission of data. Instead, asynchronous communication requires the use of a start and stop bit that identifies the beginning and end of data. As an example of asynchronous communication, both parties involved in a telephone conversation can speak at any time during the conversation.

Transmitting the start bit indicates the start of each character; issuing the stop bit indicates the end of the character. Without the use of a start and stop bit, the receiving system would have difficulty separating the data from noise. Even if the modem clocks do not have exact synchronization, the data transfer remains successful. During transmission, the modems need to stay synchronized only for the length of time needed to send 8 bits of data.

Remote Access Security

Because of the prospects for network security violations, manufacturers have supplemented remote access solutions with a variety of security enhancements. At one level, the use of a remote node access server improves network security because of the limitation of access to network resources from a single point of entry. In addition, remote access requires a combination of user identification and passwords.

Some products also offer a programmed callback feature that prompts a remote server access modem to call a remote caller back from an internal, preprogrammed list of preauthorized telephone numbers or to a user-defined telephone number. Several products use dialback pass-through — an option that establishes a dialback dependent on whether the line in use accepts incoming calls.

More exotic security options involve the use of tokenized security technologies that rely on software at the server and a smart card and personal identification number held by the user. Unique to each remote user, the smart card generates a random code that changes periodically. In turn, the server software recognizes and verifies the code and grants access to the network. Other devices include a Remote Security Attachment, which connects to the output port of the remote computer and assigns a unique network identification to the computer and the use of an authentication server that responds with an encrypted validation.

AccessBuilder by 3Com permits network managers to manage security centrally through a security server. Residing on the network node, a software agent serves between AccessBuilder and the network security server. The software agent emulates the security server and performs authentication activities. With a central database, network managers can authenticate the identity of a user, confirm access to specific corporate data, and maintain a record of login events. The AccessBuilder software notifies a network administrator about any unauthorized attempts to access the network.

Remote Access Management

Management of remote access server activity involves the capability to view all remote node servers, set up ports through a graphical port image, track connections and length of connections, monitor network errors, and monitor usage based on network protocols. Most remote access management packages provide the capability to configure, monitor, and troubleshoot the remote access system from any computer connected to the network. In addition, remote access management restricts the use of a single-access server as a primary entry point for all remote users and organizes all network users into logical groupings.

Based on SNMP, remote access management software generates packet and client use statistics, provides an overview of port status, and maintains trace and audit logs. Additionally, the software tracks connection costs in terms of bandwidth usage. For LAN-to-LAN connections, the software has a bandwidth-on-demand feature that adds more throughput as required by a particular connection.

Facsimile Servers

Facsimile transmissions have become a common and sometimes indispensable part of business transactions and communication. Because of several technological advances, the facsimile, computing, and networking worlds have converged. Network fax servers offer benefits that include the centralization of management capabilities, the reduction of long-distance telephone bills, and the elimination of routines associated with traditional fax machines. Through the use of a facsimile server, an organization can fax a document directly from a LAN-connected microcomputer without the need to print the document.

Management techniques employed with facsimile servers include providing current and long-term status information about inbound and outbound faxes. The accumulated data provides information about the successful transmission of faxes, the sending time of the faxes, the number of pages sent, and the number of retries required. Some systems also offer an account transaction module that tracks fax usage and generates chargeback and billing reports. In addition, centralized management systems may allow changes to priority settings that may affect the order of jobs in the queue.

As already mentioned, the sending of facsimiles through a local-area network takes advantage of several different technologies. The process requires a dedicated workstation equipped with fax server software and a fax board, but it also requires a printer that includes facsimile hardware. The file/fax server combines the operating characteristics of a file server with fax server software and a fax board. In addition, a standalone, dedicated fax server unit attaches directly to the network and telephone lines.

Sending the Facsimile

Most facsimile servers transmit a fax using the same methods and protocols that support the sending of electronic mail messages. The use of the print capture utility redirects the print output from applications to the server. Fax servers usually incorporate image-processing capabilities that allow the integration of company logos, signatures, and other images into a document. Compared to traditional methods, fax servers ease the processes of broadcasting and delayed transmission of facsimile messages. Delayed transmission collects outgoing faxes in batch jobs and transmits the documents during a later time.

Receiving the Facsmile

In contrast to the straightforward sending of faxes from a facsimile server, the reception of electronic faxes offers several challenges. Because the transmission of faxes involves the sending of a document as an image file rather than as a text file, computer systems sometimes have difficulty in obtaining information from the file.

Inbound routing techniques used with facsimile servers include the manual distribution of fax documents, Direct Inward Dialing, Dual-Tone Multifrequency Dialing, Channel- or Line-Based Routing, and Source ID Routing. With manual distribution, an operator views incoming faxes and routes the document to the appropriate recipients. Direct Inward Dialing, or DID, assigns and links a fax number to the e-mail address of each user on the network. After receiving facsimile messages, the fax server routs the messages according to fax number and notifies users about the placement of the fax message in their electronic mail box. Direct Inward Dialing requires a separate trunk line, separate fax numbers for each possible recipient, and a special DID interface that converts the DID line into a regular line.

With Dual-Tone Multifrequency, DTMF, Dialing, the sender must dial the recipient's special extension after making a fax-to-fax connection. The receiving machine identifies the recipient by the tones and routes the document to the appropriate location. Tones or a voice prompt sometimes signals the sender to enter an extension number.

Channel- or Line-Based Routing assigns a separate fax line to each recipient or department within an organization. With the lines feeding into the fax server, the server receives all incoming faxes and routs the documents to the appropriate user or department. The Source ID Routing method routs faxes according to the origination point. To rout in this way, Source ID Routing maintains a list of fax numbers and forwards the incoming faxes to a recipient designated to receive all faxes for a particular number.

CD-ROM and DVD Servers

A CD-ROM or DVD server makes a number of CD-ROMs or DVDs available to network users. The use of a CD-ROM or DVD server can ease the workload of system administrators through the capability to install software over the network from the server. As pictured in figure 5.4, CD-ROM or DVD servers provide shared access to databases, documentation, and other software. With optical disk servers becoming part of the network domain or NDS tree, a system administrator can grant and revoke permissions to the servers.

Most optical disk servers can interoperate with a number of network operating systems and protocols. Along with the capability to share data with network clients, most optical disk servers can share data as NFS mountable volumes. In addition, system administrators can configure the servers over the Web.

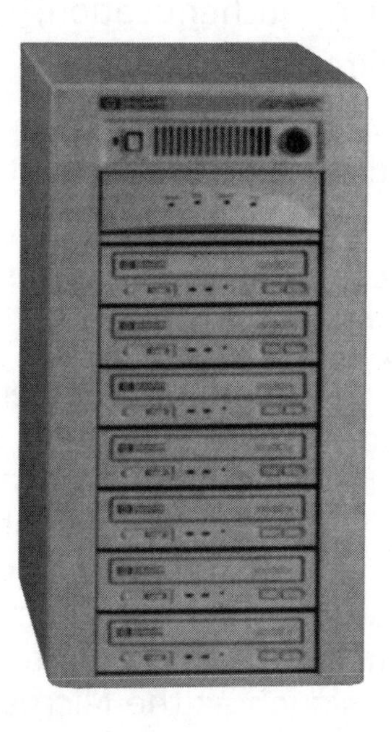

Figure 5.4. A CD-ROM server

Despite the advantages offered by optical disk servers, several disadvantages also exist. Even though standards exist, some manufacturers may not allow the sharing of optical-based information under some configurations. In addition, the use of multimedia software on CD-ROMs and DVDs also offers special problems for network users because no method exists for sharing the audio portion of the software over a network. Although compact disk technologies offer good data transfer rates, the drives have very slow random-access speeds.

To lessen the problems with speed, most optical disk servers cache as much as 160 Mb of data to the controller memory.

Some manufacturers provide a hard disk as a caching solution. In addition, many CD-ROM and DVD servers allow system administrators to create an image of the disks onto the internal hard disk of the server.

Web Servers

A Web server delivers hypertext documents through the HyperText Transfer Protocol, or HTTP, standard. The different varieties of available Web servers include, among others, the following features:

- user authentication,
- directory indexing,
- search capabilities,
- directory level security,
- virtual hosting,
- customized responses to errors, and
- user directories.

In addition, Web server tools may also provide the capability to create content and manage site functionality. Many Web servers support Java and JavaScript for establishing server-side and client-side content and Application Programming Interfaces for establishing program interaction. Active Server Pages, or ASP, allow scripted pages residing on the server to generate dynamic content when requested by a client browser. Tools such as the Microsoft ActiveX Data Objects, or ADO, allow a connection from the Web server to a database.

Accessing textual, video, or audio content through the World Wide Web involves communication between a Web-browser client and a Web server that utilizes the HyperText Transport Protocol. Because the content follows the path shown in figure 5.5, all communication occurs through the Internet. The communication begins with the typing of a URL into the

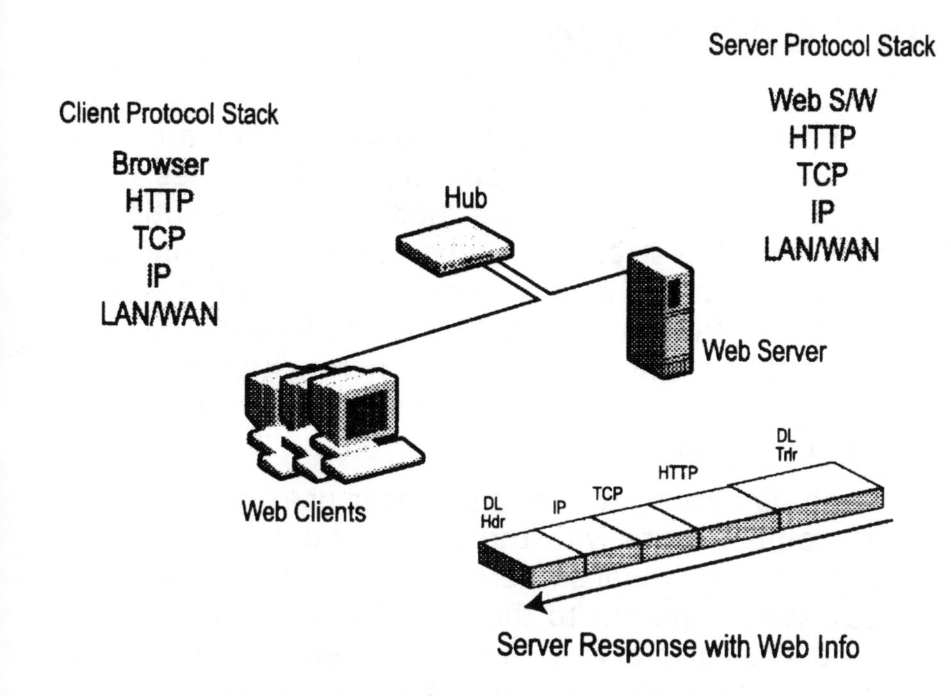

Figure 5.5. Diagram of a Web server

Web browser. Then, the client computer locates the IP address associated with the domain name given for the primary Domain Name System, or DNS, server.

As soon as the client computer and the browser begin to request and receive files from the same URL, the constant need for a DNS lookup does not exist. When the client computer accesses a different URL and begins to retrieve a different Web site, the DNS lookup initiates. Web servers may exist in any facility that has Internet connectivity.

Web Server Farms

Web server farms offer the advantage of replicated disk content. However, the use of replicated disks can result in higher expenditures. The reliance on replicated content requires content synchronization, or the duplication of any change to content data on all nodes.

Load Balancing

As depicted in figure 5.6, load balancing for a cluster of Web servers distributes the load across the nodes equally. Several hardware and software load balancer solutions can distribute incoming stream of requests among a group of Web servers. Hardware load balancers operate between the Internet and Web server farm by connecting to the Internet router and the internal LAN using two separate network segments. The load balancer acts as a fast regulating valve between the Internet and the pool of servers.

During operation, the load balancer uses a virtual IP address to communicate with the router and masks the IP addresses of the individual servers. Because only the virtual address becomes apparent to the Internet community, the load balancer acts as a safety net. The other network segment connects to a hub or switch with a pool of multiple physical servers attached.

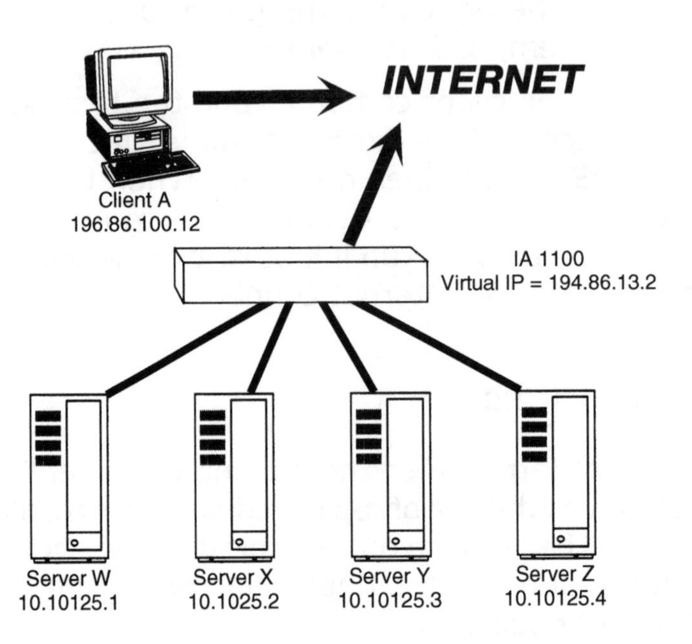

Figure 5.6. Server load balancing

Balancing methods across different switches vary. However, most methods forward the request to the least loaded server in a cluster. Although the use of only one switch introduces a single point of failure, the minimal configuration of two switches increases costs. In every instance, load balancing across switches can introduce the opportunity for bottlenecks to occur.

Software load balancing on a cluster uses the Domain Name System server. Round-Robin DNS — built in to the newest version of DNS — distributes the access among the nodes in the cluster. For name resolution, Round-Robin DNS returns the IP address list of nodes in a cluster and places the different address first in the list for each successive hardware switch.

Another load-balancing method statically partitions and assigns customers to the servers. As an example, load balancing would partition 100 customers as 10 customers per server within a configuration consisting of 10 Web servers. However, static partitioning does not account for changing traffic patterns or changes in the content of the sites. The designated partitions cannot adjust to accommodate any changes in traffic or site dynamics.

Proxy Servers

A proxy server provides filtering and caching for a local-area or enterprise network. With those two functions, proxy servers improve the performance and security at the point where a local-area network interconnects with a wide-area network or the Internet. Filtering provides a predefined collection of sites that users can visit. Caching allows frequently accessed Web documents or Web sites to store on the proxy server. Rather than going on to the Internet to download a page, a user can access a proxy server cache for the site and gain faster response times.

Most proxy servers consist of dedicated Pentium-based computer systems that run only the proxy server software and have approximately 10 megabytes of RAM per user. System administrators usually place proxy servers at the point of possible bottlenecks. Proxy servers can typically handle from 10 to 50 simultaneous users. The use of multiple proxy servers often requires the application of a load balancing mechanism.

6

Ethernet Technologies

Introduction

The different Ethernet technologies described in this chapter began as a concept that allocated the use of a shared channel. Another concept subdivided a transmission into frames of data. As the concepts moved to reality, engineering teams developed channel contention systems that eventually formed the basis for the Ethernet technologies. Subdividing a transmission into frames of data served as the basis for packet-switching networks. The importance of both concepts becomes more than evident throughout this chapter.

During the mid-1970s, the Xerox Corporation, the Intel Corporation, and the Digital Equipment Corporation combined their resources to develop Ethernet specifications. The term "Ethernet" refers to the family of baseband local area network implementations that includes:

- Ethernet networks operating at 10 Mbps,
- 100-Mbps Fast Ethernet networks, and
- Gigabit Ethernet networks operating at 1000 Mbps.

Ethernet

The Ethernet standard offers a combination of tremendous flexibility and relative simplicity in terms of implementation and understanding. An Ethernet network operates as a packet-based network built around the Carrier Sense Multiple Access with Collision Detect, or CSMA/CD, protocol. With the use of the CSMA/CD Protocol, any Ethernet node determines if it can transmit over a shared medium. The MAC layer enforces the protocol.

In brief, an Ethernet system consists of

- the physical medium used to carry Ethernet signals between computers,
- a set of medium access control rules embedded in each Ethernet interface that
- allow multiple computers to arbitrate access to the shared Ethernet channel, and
- an Ethernet frame that consists of a standardized set of bits used to carry data over the system.

Ethernet networks transmit data over twisted-pair, coaxial cable, or wireless transmission media. However, even though the network interface card supports the Ethernet standard through the use of a transceiver that performs Physical Layer functions, cables or wireless transmissions connect the nodes to the networks. We can refer to the connecting cable as an attachment unit interface, or AUI, and the transceiver as a Media Access Unit, or MAU.

An MAU covers the responsibility for connecting physically and electrically to and from a medium. In contrast, the AUI provides a defined interface to connect the Physical Signaling function of the Ethernet to the MAU. The Physical Signaling function and the AUI support signals traveling between the MAC layer and the MAU. An AUI interface may consist of a special cable and connector assembly but usually consists of an integrated circuit.

In the Ethernet architecture, devices connect to a shared medium and have equal priority access to the medium. All Ethernet nodes have permission to receive network traffic. However, only one device can transmit at any time. Because carrier sensing involves sensing the activity on the medium, the node must wait until activity on the medium has ceased before initiating a transmission. Collision detection occurs if the data transmitted by the first node collides with the data transmitted by a second node.

CSMA/CD Protocol

Referring to the OSI and CSMA/CD model shown in figure 6.1, Ethernet networks use the Carrier Sense Multiple Access/ Collision Detect Protocol for carrier transmission access. In an Ethernet network, multiple access means that any device may attempt to send a frame at any time. Each device senses the condition of the line. If the line has an idle condition, the device begins to transmit its first frame. If another device attempts to transmit data at the same time, a collision occurs.

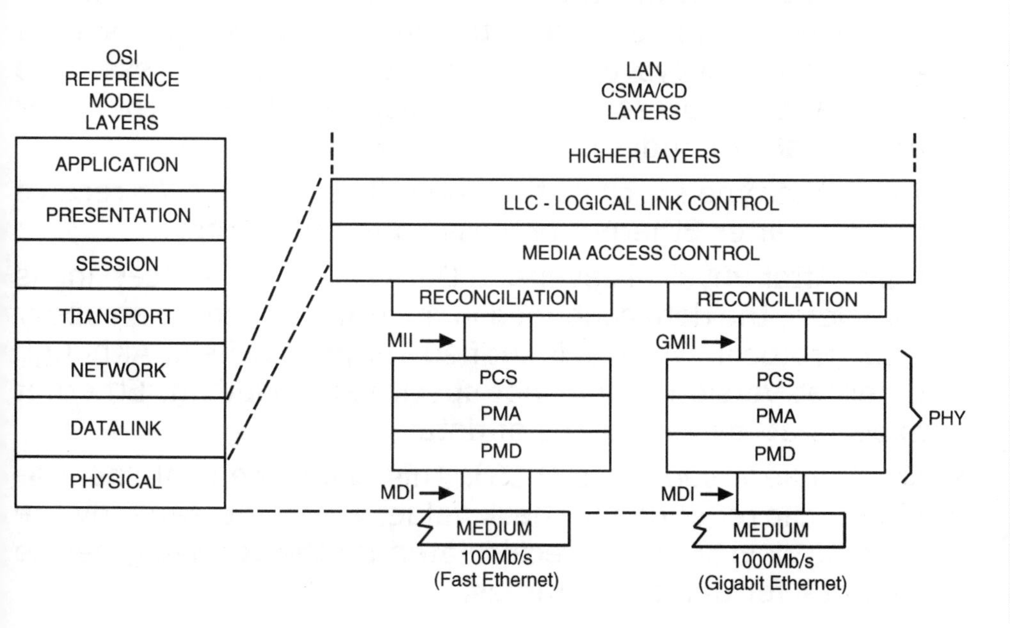

Figure 6.1. OSI and CSMACD models

Collision detect occurs, and the network discards the frames. In addition, each device waits a random amount of time and retries until successfully transmitting the data.

With the broadcast-based environment found with Ethernet networks, all stations see all frames placed on the network. Following any transmission, each station must examine every frame to determine whether the destination of the frame corresponds with the station. Frames intended for a given station pass to a higher layer protocol.

The Ethernet frame fields include the following information:

- *Preamble.* The alternating pattern of ones and zeros tells receiving stations that a frame is in transit. The Ethernet frame includes an additional byte that is the equivalent of the Start-of-Frame field.
- *Start-of-frame (SOF).* The delimiter byte ends with two consecutive 1 bits, which serve to synchronize the frame-reception portions of all stations on the LAN. SOF is explicitly specified in Ethernet.
- *Destination and source addresses.* On a vendor-dependent basis, IEEE specifies the first 3 bytes of the addresses. The Ethernet vendor specifies the last 3 bytes. The source address is always a unicast (single-node) address. The destination address can be unicast, multicast (group), or broadcast (all nodes).
- *Type.* The type specifies the upper layer protocol to receive the data after Ethernet processing is completed.
- *Data.* After Physical Layer and Data Link Layer processing is complete, the data contained in the frame is sent to an upper layer protocol, which is identified in the Type field. Although Ethernet Version 2 does not specify any padding, Ethernet expects at least 46 bytes of data.
- *Frame check sequence (FCS).* This sequence contains a 4-byte cyclic redundancy check value, which is created by the sending device and is recalculated by the receiving device to check for damaged frames.

CSMA/CD and Propagation Delay

The propagation delay of a network based on CSMA/CD equals the time needed for a signal to travel between the two furthermost stations on the LAN. As a total, the delay includes the time to pass through the lengths of media, any repeaters, and the transceivers. Media delay usually equals approximately 5 microseconds per kilometer ms/km.

CSMA/CD Collisions

Before a CSMA/CD station begins to transmit, it listens through its receiver in a state called "listen before talk," or LBT. If the network remains quiet, the station may transmit. A collision occurs when a station has just begun to transmit and one or more stations listen but do not hear the first station because of propagation delay. When the other stations begin to transmit, two or more stations transmit concurrently and the probability of a data collision exists. The probability has a proportional relationship to the number of stations, the number of transmissions per interval, the length of the network, the utilization of the network, and the size of the collision windows.

A collision window is the time when a collision may occur. The maximum time for a collision window equals the time needed for the signal to travel the path it must follow between two distant stations in a maximum length CSMA/CD LAN. Window size for any particular collision situation equals the longest delay between all stations transmitting concurrently.

Adding load to a network raises LAN utilization and gives stations less idle time to complete transmissions. Increasing the number of collision windows by using more transmissions to send the same amount of data increases overhead and utilization. Higher utilization increases the probability of multiple stations concurrently hearing a signal when beginning to transmit and causes the stations to transmit data simultaneously. Increasing the length of the corresponding collision windows increases the probability of collisions for all involved stations. Collision window size on a 10-Mbps Ethernet network equals the one-way propagation delay between competing stations.

Collision Detection

With Listen While Talk, or LWT, a baseband transmitting station listens during the transmission to determine whether another station has started to transmit. LWT forms the basis for baseband collision detection — a process accomplished by detecting that the magnitude of the signal on the bus has increased to a level higher than that achievable by the signal transmission of one station. Collision detection depends on the strength of the signal between the two furthermost stations. The signal must have a level higher than normal noise levels.

Minimum Transmit Time

The Minimum Transit Time is the minimum time that a station must transmit to ensure that it hears all possible collisions. Minimum Transit Time must equal at least twice the one-way sum of all propagation delays in a maximum length CSMA/CD network. Baseband CSMA/CD 10-Mbps networks transmit a minimum frame length of 512 bits. Due to the requirement for Minimum Transit Time, the frame length also remains constant.

Collision Resolution

CSMA/CD includes a mechanism to control collision resolution or the collisions that occur during an increased network load. However, the method cannot introduce significant retransmission. When a collision occurs, the first station to detect the collision issues a "jam" to notify all listening stations. As a result of each station delaying for a random interval before attempting to retransmit, CSMA/CD invokes a binary exponential backoff algorithm. If the next attempt of a station to transmit results in a collision, CSMA/CD increases the backoff delay for that station. The action may occur 16 times before the station reports an error.

The IEEE 802.3 Standard

Although the IEEE 802.3 standard has a basis in the original Ethernet concept, the standard support of multiple Physical Layer options, higher data transmission rates, and different signaling methods. With the support of multiple Physical Layer options, the 802.3 standard utilizes 50- and 75-ohm coaxial cable, unshielded twisted-pair cable, and fiber-optic cable. In addition, the 802.3 standard supports longer maximum cable segment lengths.

Difference	Description
MAC Layer Frame Format	IEEE 802.3 includes a length field. The original specification requires a higher layer protocol to pad the frame and a type field to specify the client protocol.
Signal Quality Error	IEEE 802.3 specifies an SQE message sent by the MAU to determine the quality of the line and to indicate if a collision has occurred. The signal verifies the operation of the collision signaling function at the end of the transmission. The signal also monitors the line for improper signals and breaks in the circuit.
Repeaters and Transceivers	With the original Ethernet standard, a maximum of two repeaters can operate between any two MAUs. With IEEE 802.3, a repeater requires a transceiver connection on both of the segments that it joins and counts toward a 100 transceiver limit for each segment.
Transmission Speed	The original Ethernet supported only a 10-Mbps data rate. Even though the original 802.3 covered data rates ranging from 1 to 20 Mbps, newer IEEE specifications have increased the range.

TABLE 6.1 — COMPARISON OF ORIGINAL ETHERNET AND IEEE 802.3 DIFFERENCES

Although the IEEE 802.3 standard includes much of the original Ethernet specifications, table 6.1 shows that several differences exist. As the table indicates, the IEEE improvements covered the MAC-layer frame format, the use of the signal-quality-error, or SQE, message, and the use of repeaters and transceivers.

Referring to figure 6.2, an Ethernet network relies on MAC layer functions to pass data frames from station to station. The data frames consist of data bits grouped within the specified MAC frame format. Each packet begins with a preamble sequence that uses an alternating one/zero pattern to produce a single 5-MHz frequency on the network. With Manchester encoding and decoding in place, the use of the frequency at the start of each packet allows the receiver to lock onto the incoming bit stream.

Figure 6.3 compares the MAC frames for the original Ethernet packet format and the IEEE 802.3 packet format and illustrates the MAC layer differences between the two standards. The Ethernet standard uses a synch field to indicate that the data portion of the message follows the preamble; the 802.3 standard uses a Start Frame Delimiter for the same purpose. The MAC layer at the receiving station uses the Des-

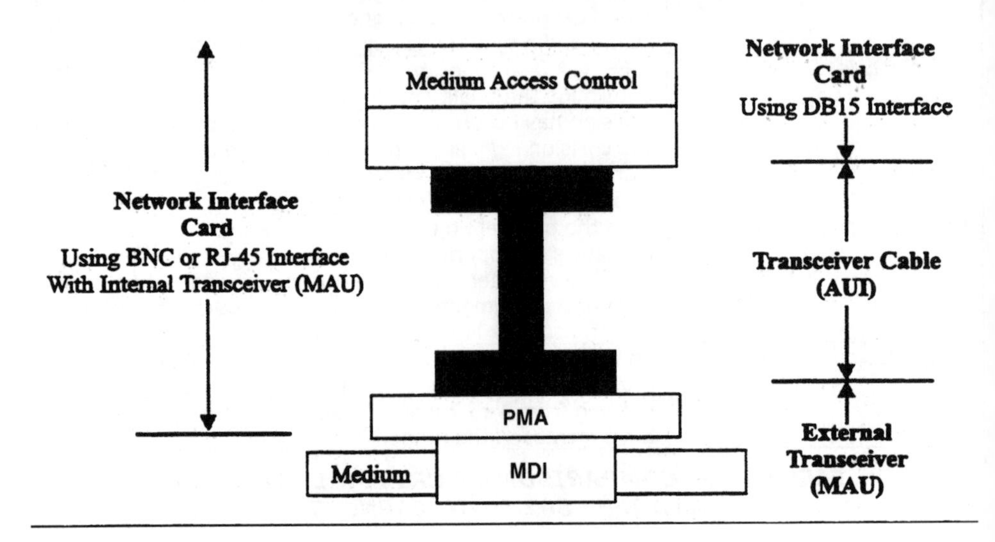

Figure 6.2. IEEE 802.3 Ethernet media attachment

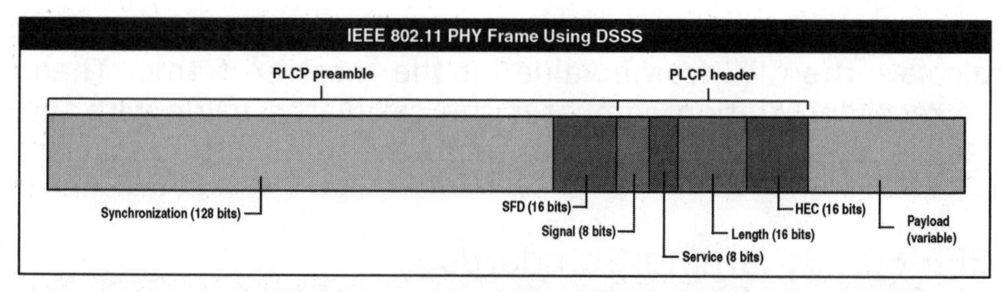

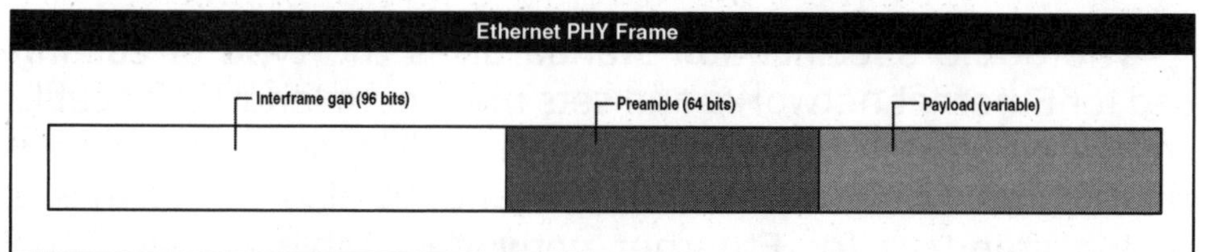

Figure 6.3. Comparison of the 802.3 and Ethernet MAC layer frames

tination Address to determine if the address of the incoming packet refers to a particular node. Both the Ethernet and the 802.3 standard support individual and unique addresses, multicast and group addresses, and broadcast messages within the Destination Address.

The MAC layer at the transmitting station inserts its own unique address into the Source Address as an indicator about the originating station. Depending on the use of either the Ethernet or 802.3 specifications, either a Length or Type field follows the Source Address field. The Ethernet standard uses a 2-byte Type field, and the 802.3 standard uses a 2-byte Length field.

As mentioned in table 6.1, the 802.3 specifications append pad characters to the LLC Data field before sending data over the network. The 802.3 standard allows a slightly smaller data frame than the value seen with the Ethernet standard. In comparison, the Ethernet standard ensures that a minimum data field of 46 bytes exists before data passes to the MAC layer.

The FCS contains the cyclic redundancy check for the entire frame. While the transmitting station calculates the CRC on the Destination Address, Source Address, Length/Type, and Data fields and appends the CRC to the last four bytes of the frame, the receiving station uses the same CRC algorithm to calculate the CRC frame value for the received frame. Then, the receiving station compares the calculated value with the CRC value appended to the frame by the transmitting station.

IEEE 802.3 Naming Standards

The 802.3 specification standardizes the type of cabling used for Ethernet networks and sets maximum lengths for cable runs between repeaters. Along with adding support for new Ethernet technologies, the 802.3 standard also establishes a naming standard for Ethernet networks. Table 6.2 lists the naming standards.

Standard	Media/Topology
10Base2	Thinnet
10Base5	Thicknet
1Base5	StarLAN
10Broad36	Broadband — CATV
10BaseT	Twisted-pair
10BaseF	Fiber optic
100BaseT4	Twisted-pair
100BaseTX	Twisted-pair
100BaseFX	Fiber optic

TABLE 6.2 — IEEE 802.3 NAMING STANDARDS

As illustrated in figure 6.4, the format offers the "s type 1" layout with the following designations:

- "s" for speed in megabits per second,
- "type" for either baseband or broadband communication, and
- "1" for the maximum segment length in 100-meter intervals.

As an example, 10Base5 specifies 10 Mbps in a baseband communication across a segment that has a maximum length of 500 meters.

However, exceptions exist for different types of Ethernet networks. If we break 10BaseT into parts, the 10 represents the 10-Mbps transmission rate while the "base" signifies baseband communication. The "T" shows that the standard

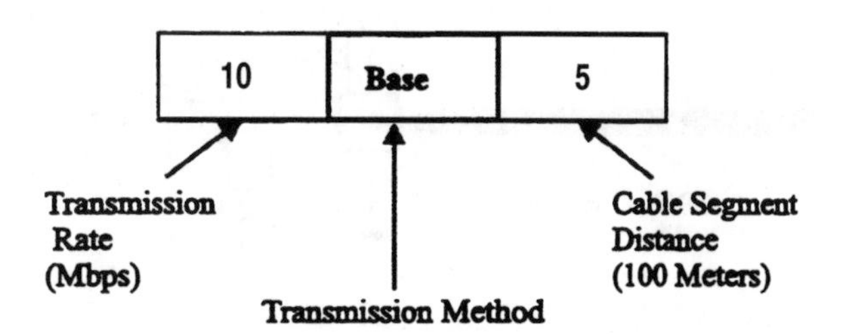

Figure 6.4. Illustration of the Ethernet naming standard

works through twisted-pair cabling. Because the 10BaseT standard uses baseband communication, one cabling pair carries transmitted data, and the other pair carries received data.

Classifying Ethernet Networks

Figure 6.5 shows the most basic Ethernet network configuration. In the figure, several workstations connect together on a single Ethernet segment. Based on the 10Base2 standard, the network includes four network nodes and a single server. This particular network relies on Thinnet coaxial cabling and a network interface card that contains the transceiver.

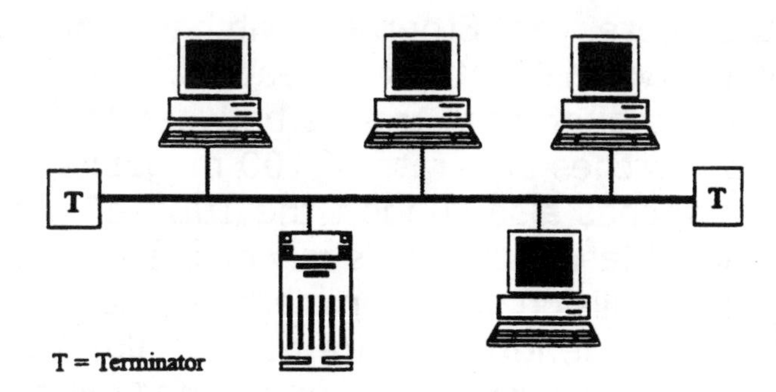

Figure 6.5. Basic Ethernet bus topology

The 10Base2 Ethernet network shown in figure 6.6 relies on Thinnet coaxial cabling and external transceivers. A transceiver takes the digital signal found at the workstation and converts the signal into the format needed for physical cabling. Each segment of the network terminates at both ends.

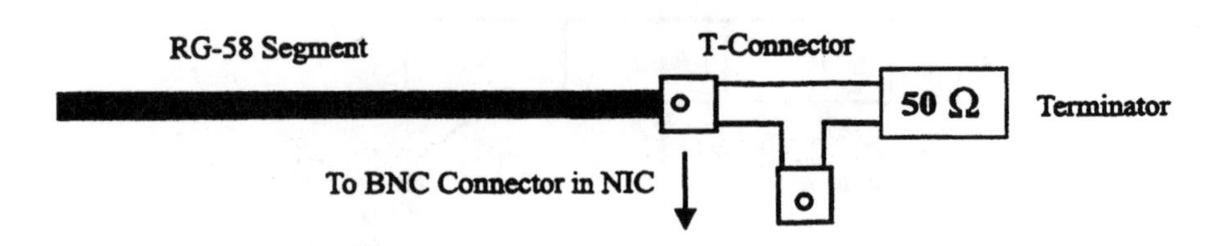

Figure 6.6. 10Base2 network

Category 3 and Category 5 twisted-pair cabling supports the 10BaseT data transmission standard. The 10BaseT over Category 3 cable has a maximum length of 100 meters, but the use of Category 5 cabling for the 10BaseT standard increases the cable length to 150 meters. One cable contains the two pair of wires and may bundle multiple pairs together. Each end of the 10BaseT pair set terminates with an eight-position jack.

Fast Ethernet Networks

As depicted in figure 6.7, Fast Ethernet provides a 100-Mbps high-speed LAN technology that offers increased bandwidth. In the figure, Fast Ethernet switches connect between server farms, two 100 Mbps repeaters, and two 10/100 switches. Even though data travels between the servers, repeaters, and switches at a rate of 100 megabits per second, the 10/100 switches also support the 10BaseT network. The 100BaseT and 10BaseT standards rely on the same IEEE 802.3 MAC access and collision detection methods and have the same frame format and length requirements. Like 10BaseT, the 100BaseT standard operates over unshielded twisted-pair and shielded twisted-pair cabling.

The 100BaseT Ethernet Standard relies on Category 5 twisted-pair cabling and carries a signal at a data rate of 100 Mbps. Applications for 100BaseT over Cat 5 cabling have a maximum cable length of 100 meters. In comparison to the 100BaseT standard, 100BaseT2 supports a data transmission

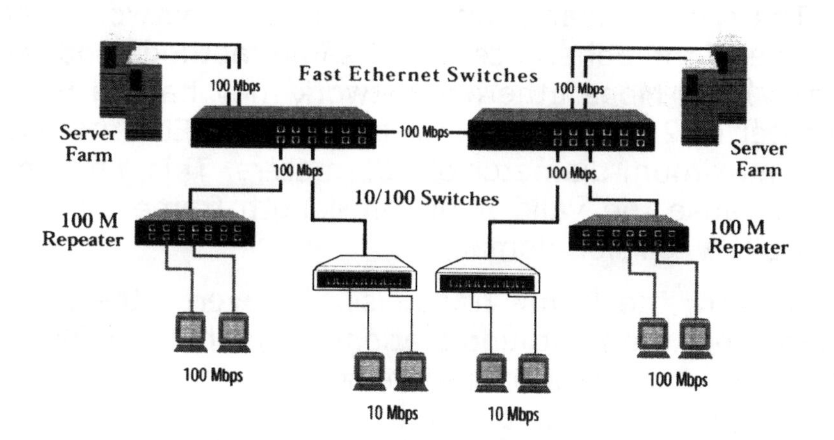

Figure 6.7. Switched Fast Ethernet network

rate of 100 Mbps over Category 3 twisted-pair cabling rather than requiring the use of Cat 5 cabling.

In addition, 100BaseT supports all applications and networking software currently running on 802.3 networks and supports dual speeds of 10 and 100 Mbps using 100BaseT fast link pulses. As a result, Fast Ethernet hubs and adapter cards detect and support 10 and 100-Mbps data transfer rates. Figure 6.8 illustrates how the 802.3 MAC sublayer and higher layers run unchanged on 100BaseT.

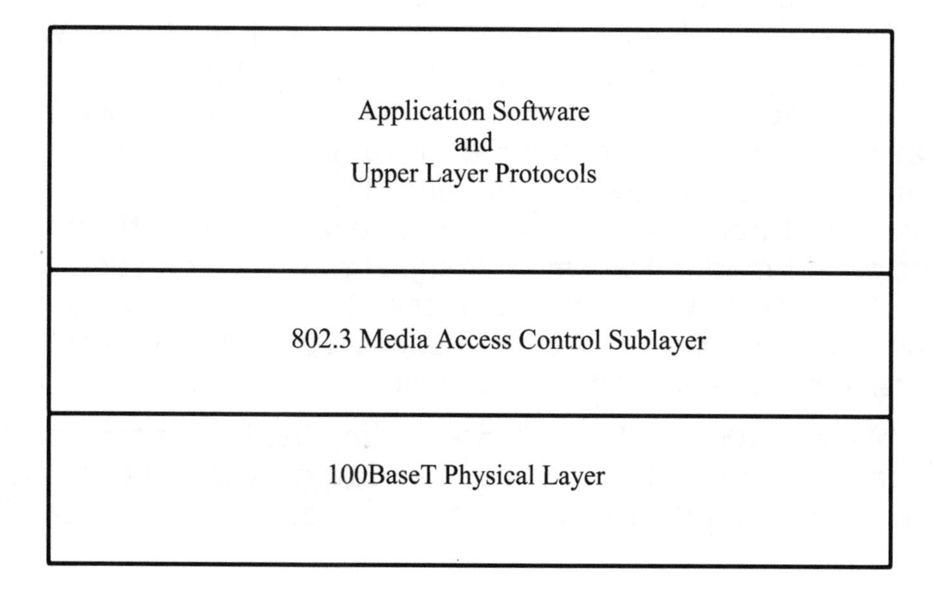

Figure 6.8. MAC layer and 100BaseT

The two standards differ in terms of network diameter because of the adherence to the same methods and requirements. A 10-Mbps Ethernet network may have a maximum diameter of 2100 meters, but a 100-Mbps Ethernet network has a maximum diameter of 205 meters. This difference occurs because the valid minimum-length frame transmission defines the collision domain.

In turn, the frame transmission governs the maximum distance between two end stations on a shared segment. As the speed of network operation increases, both the minimum frame transmission time and the maximum diameter of a collision domain decrease. The bit budget of a collision domain consists of the maximum signal delay time of the various networking components, the MAC layer of the station, and the physical medium.

The increased data transfer rate seen with 100 Mbps combines with the physical media propagation speed to limit the diameter by ten times. A station transmitting on a 100-Mbps network at ten times the speed of a station transmitting on a 10-Mbps network must have a maximum distance that measures ten times less.

The 100BaseT2 standard uses two pairs of twisted-pair cabling and a dual duplex baseband transmission scheme. With dual duplex baseband transmission, data simultaneously transmit over each wire pair in each direction. Dual duplex baseband transmission uses the complex Five-level Pulse Amplitude Modulation, or PAM5x5, signal-encoding scheme that transmits data through a quinary, or five-level signal. With this, 4 bits of information transmit per signal transition on each wire pair. As a result, the combination of a 25-megabaud-transmission rate and two pairs of cabling support the full-duplex transmission of data at a rate of 100 Mbps.

As with 100BaseT2, 100BaseT4 supports the transmission of data at a rate of 100 Mbps over Category 3 or better cabling. With 100BaseT4, however, four pairs of cabling carry the signal. One pair carries transmitted data while another pair remains dedicated to carrying received data. Two bidirectional pairs carry or receive data. By employing four cabling pairs, one dedicated pair is used for collision detection on the network while the three remaining pairs remain available for carrying the data transfer.

Data transmission over a 100BaseT4 network occurs through an "8B6T" signal-encoding scheme that converts 8 bits of binary data into six "ternary" signals. During the transmission of data over twisted-pair wires, the ternary signal may have one of three binary values. As a result, the encoding scheme splits the 100-Mbps data transmission rate over three twisted pairs with each pair carrying a 33.3-Mbps transmission rate. Ternary signaling requires only 6 bauds to transfer 8 bits of information and provides a maximum signal transmission rate of 25 megabaud on each of the twisted pairs. Since 25 megabauds translates into a maximum frequency of 12.5 MHz, the data transmission rate falls within the 16 MHz limit supported by Category 3 cabling.

Gigabit Ethernet Networks

Large networks feature a backbone or the portion of the network that carries the most significant traffic. In addition, the backbone uses a bridge to connect LANs or subnetworks together to form the larger network. As network applications have grown from using 10BaseT standards to relying on the higher bandwidth and faster data transfer speeds seen with 100Mbps Ethernet networks, the need for high-speed backbone technologies has become more apparent.

As illustrated in figure 6.9, the 1000BaseT standard — sometimes called Gigabit Ethernet — supports the transmission of data at a rate of 1000 Mbps over Category 5 balanced copper cabling. With the backbone created in figure 6.10, the

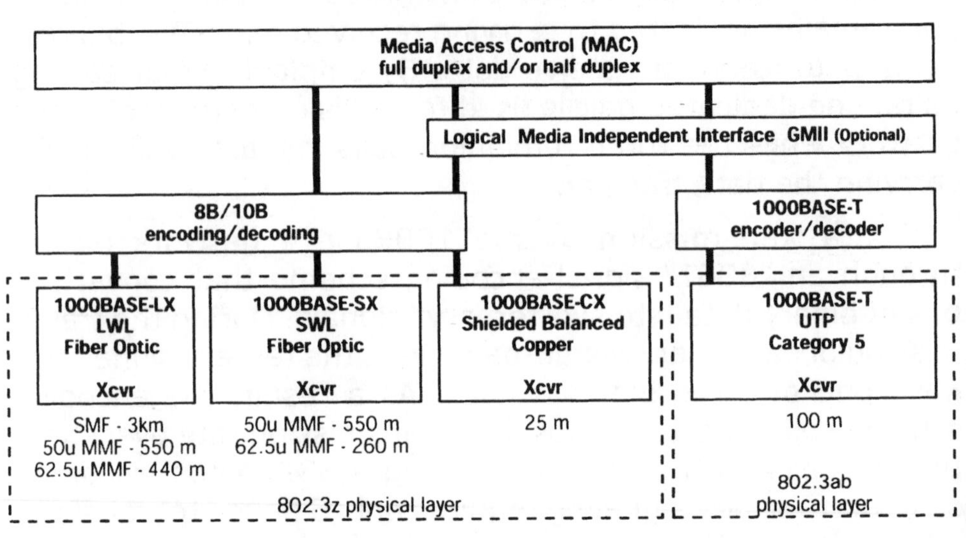

Figure 6.9. Gigabit Ethernet technologies and the MAC layers

cabling has a maximum length of 100 meters and provides full-duplex baseband transmission over four pairs. Each cabling pair carries data at a transmission rate of 250 Mbps and supports baseband signaling at a modulation rate of 125 MHz. Gigabit Ethernet relies on PAM5 encoding for the transmission of the data over each cabling pair.

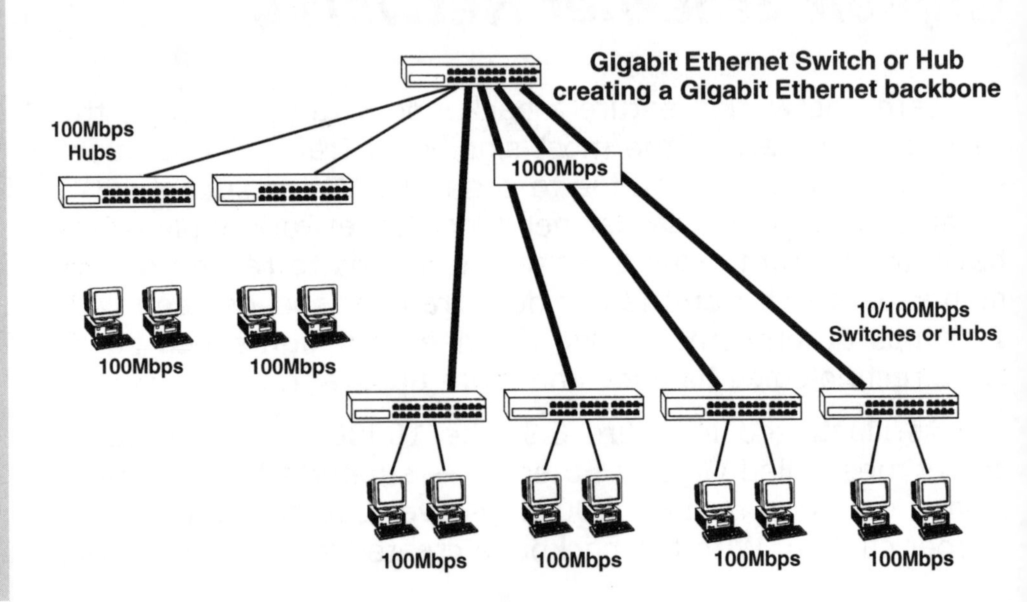

Figure 6.10. Creating a Gigabit Ethernet backbone

Gigabit Ethernet Characteristics

Due to the need for compatibility between networks and the need to provide the high-speed transport of data, the Gigabit Ethernet Task Force set a list of criteria for the new Ethernet standard. In addition to adopting the 802.3 frame format and running at 1000 Mbps, the task force saw that the new standard should

- maintain the IEEE 802 functional requirement and adopt flow control based on the 802.3 standard,
- allow simple forwarding between 10, 100, and 1000 Mbps Ethernet,
- maintain the minimum and maximum frame size of the current IEEE 802.3 standard,
- provide full- and half-duplex operation,
- support star-wired topologies
- use the CSMA/CD access method with support for at least one repeater/collision domain and support a maximum collision domain diameter of 200 meters,
- provide a family of physical layer specifications that support a link distance of at least 500 meters on multimode fiber, at least 25 meters on copper with 100 meters preferred, and at least 3,000 meters on single mode fiber, and
- specify an optional Gigabit Ethernet Media Independent Interface.

Figure 6.11 illustrates the matching of Gigabit Ethernet criteria with existing network needs, and figure 6.12 presents a diagram of an upgrade from Fast Ethernet to Gigabit Ethernet. In figure 6.12, the majority of the upgrade occurs through the replacement of a Fast Ethernet switch with a Gigabit Ethernet switch.

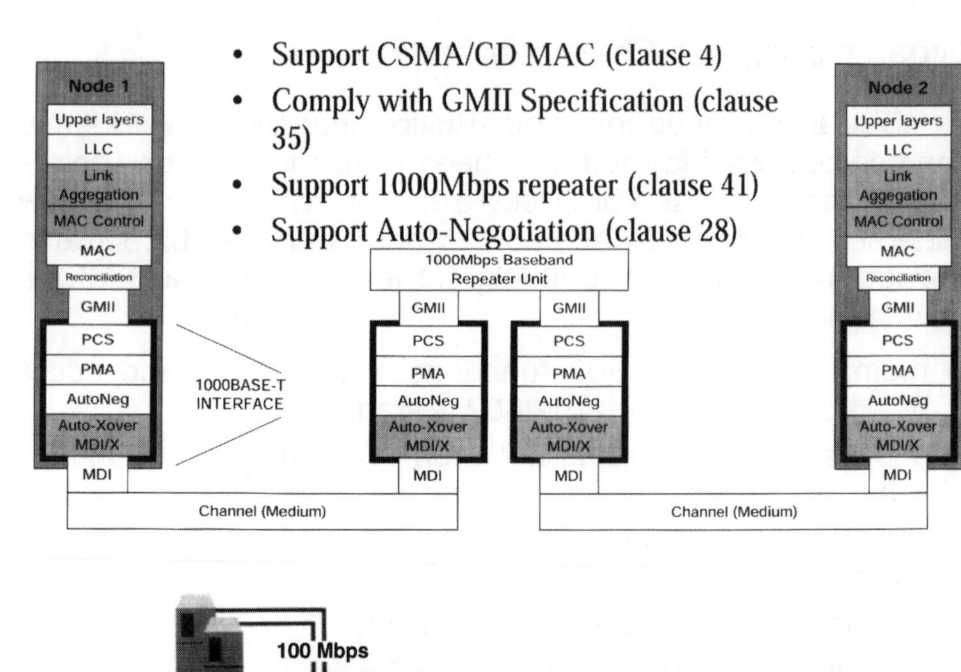

Figure 6.11. Matching Gigabit Ethernet criteria with existing needs

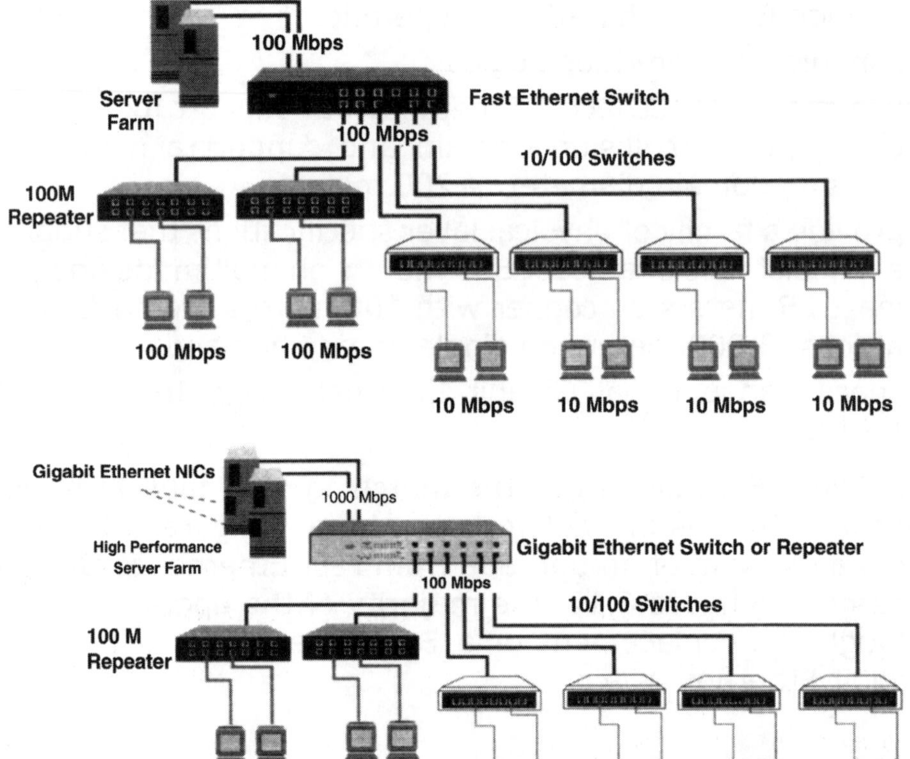

Figure 6.12. Upgrading from Fast Ethernet to Gigabit Ethernet

Compatibility

Gigabit Ethernet offers transparent compatibility with the lower speed Ethernet standards while providing high-speed data transfer speeds of 1gigabit per second (Gbps). To accelerate speeds from 100 Mbps Fast Ethernet to 1 Gbps, several changes occur at the physical interface. Because Gigabit Ethernet remains identical to 10BaseT and 100BaseT Ethernet from the Data Link Layer upward, the Gigabit Ethernet standard combines the best points of previous Ethernet standards and the Fibre Channel standard. Figure 6.13 shows the Fast Ethernet and Gigabit Ethernet CSMA/CD layers.

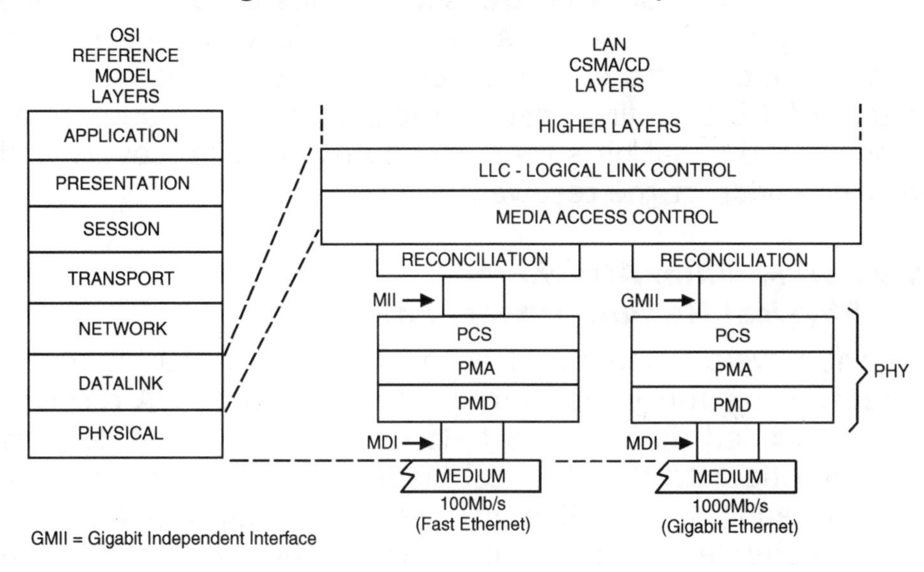

Figure 6.13. Fast Ethernet and Gigabit Ethernet CSMA/CD layers

Gigabit Physical Layer and Media

At the Physical Layer, the Gigabit Ethernet specification uses the following four forms of transmission media:

- 1000BaseLX — long-wave (LW) laser over single-mode and multimode fiber,
- 1000BaseSX — short-wave (SW) laser over multimode fiber,
- 1000BaseCX — balanced shielded 150-ohm copper cable, and
- 1000BaseT — Category 5 UTP cable.

The Gigabit Ethernet Interface Carrier, or GBIC, allows network managers to configure each Gigabit Ethernet port on a port-by-port basis for short-wave and long-wave lasers as well as for copper physical interfaces. Given the GBIC configuration, switch vendors can build a single physical switch or switch module that the customer can configure for the required laser or fiber topology.

Physical Coding Sublayer

The Physical Coding Sublayer contains the 8B/10B encoder/ decoder for use with optical or short copper links. In turn, the 8B/10B encoding scheme transmits 8 bits as a 10-bit code group. Gigabit Ethernet relies on a dc-balanced transmission code to support the electrical requirements of the receiving units. 8B/10B encoding ensures the presence of enough transitions in the serial bit stream to improve clock recovery and error detection at the receiver.

Physical Medium Attachment and Physical Medium Dependent

The Physical Medium Attachment performs 10-bit serialize/deserialize functions. During operation, the PMA receives 10-bit encoded data at 125 MHz from the PCS and delivers serialized data to the Physical Medium Dependent, or PMD, block. In addition, the PMA also receives serialized data from the PMD and delivers deserialized 10-bit data to the PCS. The PMD sublayer provides the media transceivers and connectors for various media and supports both 780- and 1300-nanometer (nm) wavelength optical drivers.

MAC Layer Characteristics

As shown in figure 6.14, the MAC layer of Gigabit Ethernet has characteristics similar to those seen with earlier Ethernet standards. As an example, the MAC Layer of Gigabit Ethernet supports both full- and half-duplex transmission. In addition,

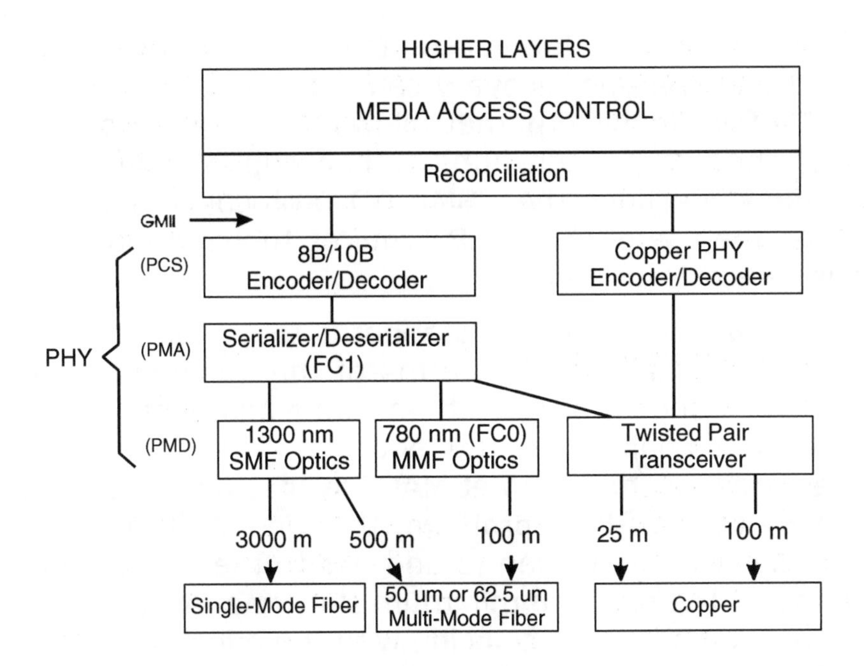

Figure 6.14. MAC layer and functional elements of Gigabit Ethernet

Gigabit Ethernet offers collision detection, maximum network diameter, and repeater rules. Gigabit Ethernet also provides support for half-duplex Ethernet while adding frame bursting and carrier extension. Because the Logical Link Control Layer remains unchanged, all upper layer protocols — such as TCP/IP — continue to support communications across the network.

Gigabit Ethernet MAC operates as a scaled-up version of the Fast Ethernet MAC with an effective data rate of 1000 Mbps. The MAC sublayer supports both full-duplex and half-duplex operation. In full-duplex mode, the MAC uses the frame-based flow control. In half-duplex mode, the MAC supports the CSMA/CD access method. The Reconciliation Sublayer and Gigabit Media Independent Interface interconnect the MAC Sublayer and Physical Layer.

Half-Duplex Gigabit Ethernet

For half-duplex transmission, Gigabit Ethernet uses the same CSMA/CD protocol as seen with other Ethernet standards to ensure that stations can communicate over a single wire and that collision recovery can occur. Implementation of CSMA/CD for Gigabit Ethernet allows the creation of shared Gigabit Ethernet through hubs or half-duplex point-to-point connections. Because the CSMA/CD protocol remains delay sensitive, the use of CSMA/CD requires the creation of a bit-budget per-collision domain.

Acceleration of Ethernet standards to gigabit speeds has created some challenges in terms of the implementation of CSMA/CD. At speeds greater than 100 Mbps, smaller packet sizes become smaller than the length of the slot-time, or the time required for the Ethernet MAC Layer to handle collisions. To remedy the slot-time problem, the Gigabit Ethernet standard uses carrier extension to add bits to the frame until the frame meets the required minimum slot-time. As a result, the smaller packet sizes can coincide with the minimum slot-time and allow seamless operation with the other Ethernet CSMA/CD standards.

The Gigabit Ethernet standard also uses frame bursting, or the transmission of a burst of frames without relinquishing control by an end station. As long as the network never appears idle, other stations defer to the frame burst. The end station using frame burst fills the interframe interval with extension bits so that the network never appears idle to other stations.

Full-Duplex Gigabit Ethernet

Most of the issues that we have considered with half-duplex Gigabit Ethernet data transmissions negate the effectiveness of the high-speed Ethernet standard. Because the use of full-duplex Ethernet eliminates collisions, full-duplex Gigabit Ethernet does not require the use of the CSMA/CD Protocol for flow control or as an access medium. As we have already discussed, full duplex provides a method for transmitting and receiving data simultaneously on a single medium.

Typical full-duplex applications occur between two end points. As an example, switch-to-switch connections, switch-to-server connections, and switch–to-router connections use full-duplex transmissions. The use of full-duplex data transmissions has easily and effectively allowed the bandwidth on Ethernet and Fast Ethernet networks to double from 10 Mbps to 20 Mbps and from 100 Mbps to 200 Mbps. Full-duplex transmission used in a Gigabit Ethernet environment promises to increase bandwidth from 1 Gbps to 2 Gbps for point-to-point links.

Gigabit Ethernet Multilayer Switching

Several methods exist for using Gigabit Ethernet to increase bandwidth and capacity within a network. Gigabit Ethernet can improve Layer 2 performance by using throughput to eliminate bottlenecks and provides an answer as the building backbone for interconnection of wiring closets. Figure 6.15 illustrates multilayer Gigabit switching designs. In this application, a Gigabit multilayer switch aggregates the traffic for the building and provides connection to servers through Gigabit Ethernet or Fast Ethernet.

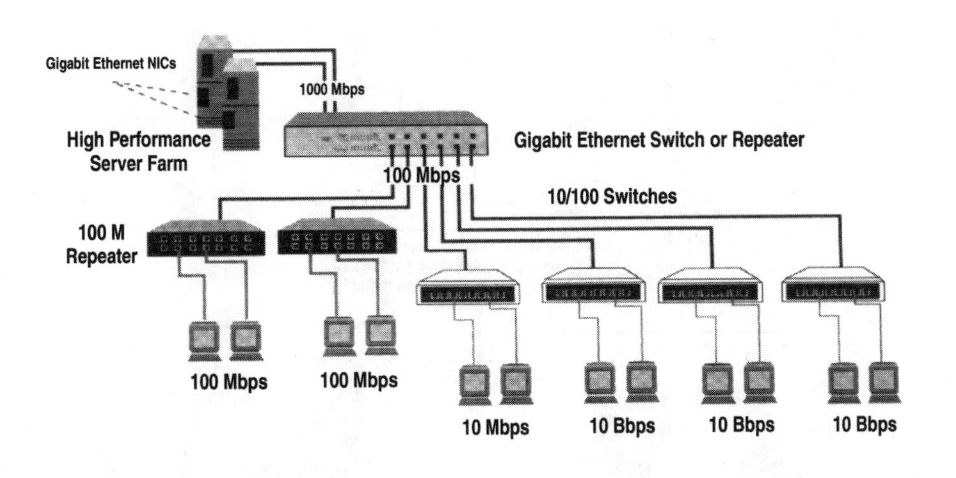

Figure 6.15. Multilayer Gigabit Ethernet switching

Shared Ethernet Configurations

Figure 6.16 shows the configuration of a 10BaseT Ethernet network. In the figure, an Ethernet hub connects eight workstations together into a star topology. The use of the hub — rather than a switch — establishes a shared domain where all nodes attached to the hub share the 10-Mbps bandwidth. UTP cabling functions as the transmission media and uses RJ-45 connectors. 10BaseT networks offer the following advantages:

- intelligent hubs,
- easily-added or removed nodes, and
- easier troubleshooting.

Hub-to-hub connections can solve the need for attaching additional nodes to the network. However, the network cannot accept any additional connections after the original hub reaches capacity. Although the second hub provides more physical ports for network connectivity, it also becomes part of the collision domain, and each additional port contends for the shared bandwidth. Therefore, the interconnected hubs will provide inadequate network performance as network traffic increases.

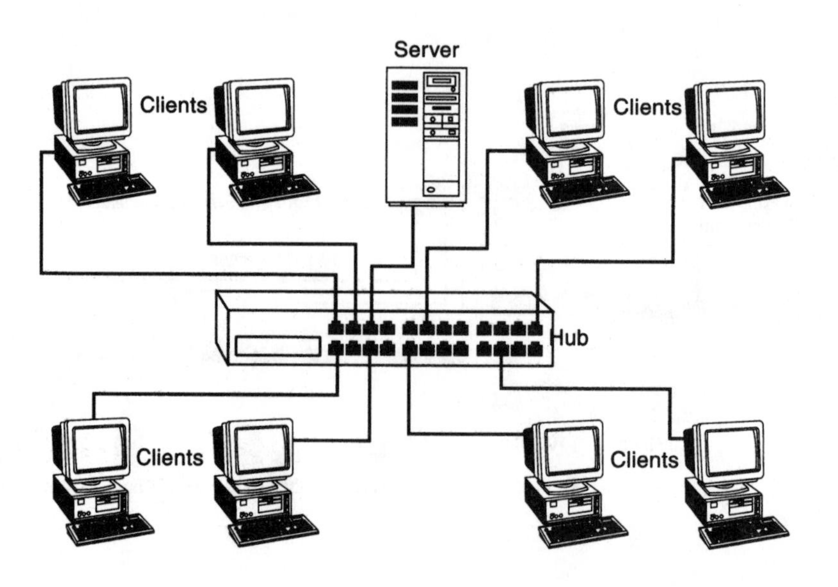

Figure 6.16. 10BaseT Ethenet network

Switched Ethernet Configurations

The network design featured in figure 6.17 solves the network performance problem by replacing the standard hub with a switching hub. With Ethernet switching, the MAC Layer address determines the switch port as the destination for the frame. In addition, a virtual connection occurs between the sending and receiving ports and prevents other ports from becoming available as destinations for the frame. As a result, the switched network does not allow data collisions to occur. The dedicated connection remains in place only for the time needed to pass the frame between the sending and receiving stations. Figure 6.18 shows a Fast Ethernet switch.

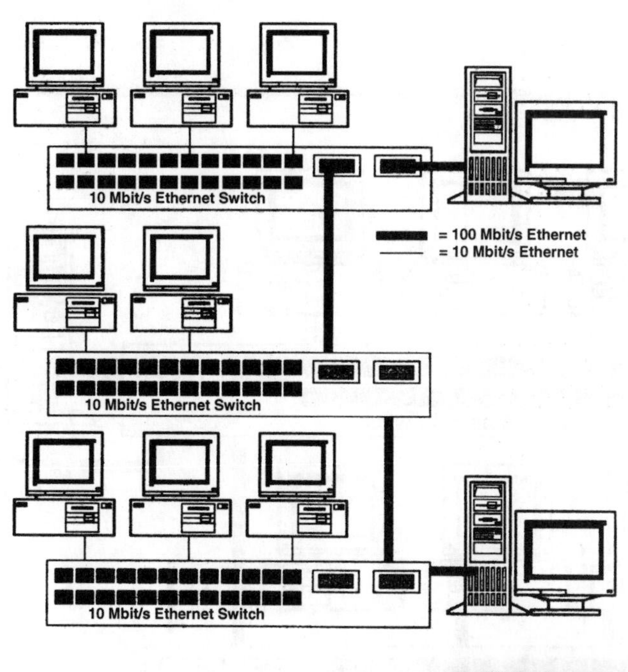

Figure 6.17. Replacing the shared hub with switched hubs

Figure 6.18. Fast Ethernet switch

If the destination port for a frame appears busy, the sending station retains the data in a buffer. Once the destination port becomes available, the station releases the frame and sends the frame to the destination. Even though using buffers works well, a buffer filled to capacity may cause the network to lose frames.

Figure 6.19 shows another network design that expands capacity by connecting hubs to a central switch. In the figure, each hub attaches to a switch port. From there, the network logically arranges users into workgroups on a hub-by-hub basis. Because the server requires access to multiple segments, the server attaches to a dedicated port on the switch. As a result, entire segments become switched, and multiple users attach to the same switch port through the same hub. Without the use of a dedicated port for the server, network performance would decrease.

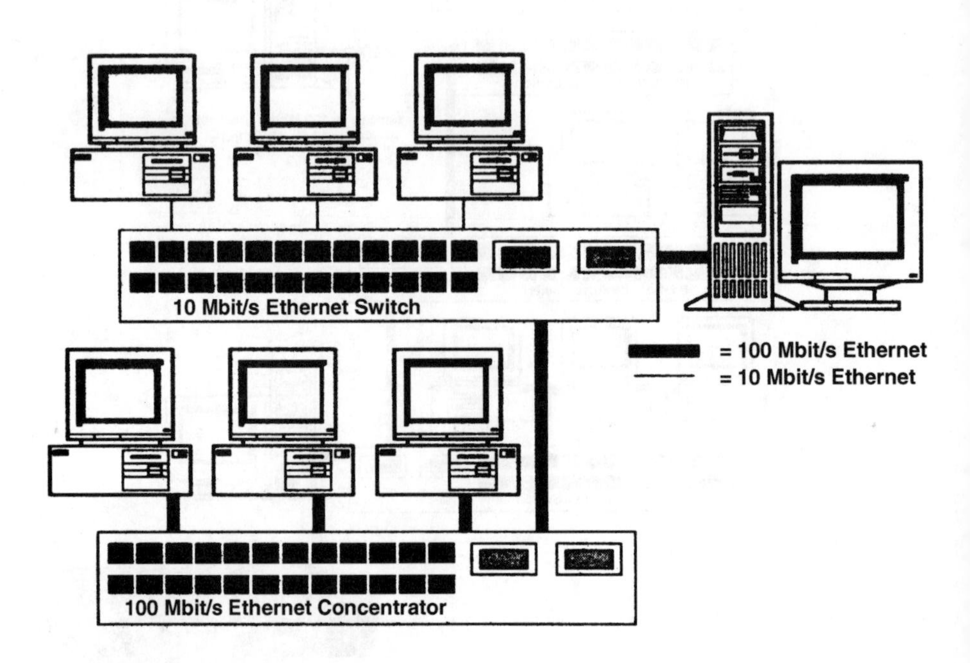

Figure 6.19. Connecting a hub to a switch

Moving back to figure 6.19, the switch offers different port speeds. As the figure shows, the server attaches to a 100-Mbps port while the connection of the workgroup hubs establishes the sharing of the 10-Mbps bandwidth. The switch does not forward frames between hubs.

When comparing the benefits of shared and switched Ethernet networks, it may become helpful to consider several differences. As an example, a shared 100-Mbps Ethernet provides a maximum data throughput of 100 Mbps across all ports of a concentrator. However, a switched 10-Mbps Ethernet network provides a dedicated 10 Mbps to each device on the switch. In addition, most Ethernet switches provide one or two high-speed uplink ports to connect to a server or to another switch. Many of these switches also support full duplex Ethernet on the 10 Mbps ports to support even higher data throughput.

A 10-Mbps Ethernet switch has a limitation of 10 Mbps per station, but a shared 100-Mbps Ethernet can establish a single connection with a maximum of 100 Mbps. The maximum throughput with shared Ethernet at the single connection occurs if no other devices contend for the bandwidth. Because of the higher bandwidth, the shared 100-Mbps Ethernet network does not have significant problems with collisions. Unlike the 10-Mbps Ethernet, the 100 Mbps Ethernet provides sufficient bandwidth to support more than 5,000 transactions per second.

A shared 100-Mbps Ethernet network usually delivers better performance in smaller networks with fewer concurrent connections. As the number of concurrent users with high bandwidth requirements grows larger, though, the switched 10-Mbps network delivers better performance. The bandwidth requirements also depend on the type of traffic across the network. With interactive traffic including most mission-critical applications, users rely on speed and consistent response times. A combination of batch and interactive traffic also affects bandwidth performance and the number of concurrent users.

Smaller networks with fewer concurrent connections and networks that have either all batch or all interactive traffic will generally experience better performance using a shared 100-Mbps Ethernet network design. The shared network can deliver more than 10 Mbps of data throughput to each user during periods of light network utilization. In most instances, a small network includes only a single server and 5 to 30 users attached to one or two switches or concentrators. If the network has both interactive and batch traffic, it could include two servers.

Larger networks that have more concurrent connections and networks that have a combination of batch and client-server or interactive traffic function better with a switched 10-Mbps Ethernet design. The switched network protects the performance of the interactive traffic while delivering better performance for batch traffic to multiple servers through increased aggregate bandwidth across multiple uplinks. A medium network usually includes two to three servers and 20 to 50 users spread across three or four switches or concentrators. Large networks include three or more servers and 50 to 150 users attached to several switches or concentrators.

An alternate solution combines Switched 10 and Shared 100 in the same network. In this case, interactive clients could connect to Switched 10 ports while the server connects to a full-duplex 100-Mbps uplink port. Batch clients connect to the shared 100 concentrator that, in turn, connects to a second uplink on the switch. Because the clients connect to 100-Mbps concentrator ports that switch between two uplink ports to the server, this configuration provides the batch clients with a full 100 Mbps of data throughput. Interactive and client-server clients connect to the switch ports. Although all batch, interactive, and client-server clients still access the single server across a common 100-Mbps uplink, this configuration delivers acceptable performance for both interactive and client-server traffic.

Cascading additional switches across the uplink ports can support more interactive and client-server users. However, users placed farther from the server have data traveling over more switch hops. As a result, uplinks closer to the server carry more traffic than uplinks found farther from the server. To conserve bandwidth, the network design requires the placement of heavy users close to the server. Because the server connection has a limit of 100 Mbps in either direction and because the connection represents a fraction of the total number of server connections, the traffic load across the uplink usually has a throughput of less than 100 Mbps.

Cascading switches through the uplink ports can provide connectivity to one or two servers; however, accessing more servers requires a more scalable design. One solution involves dedicating a separate switch or switch cascade for each server. From there, each of the switch cascades attach to a single Shared 100 concentrator.

7

Peer-to-Peer, Client-Server, and Distributed Computing

Introduction

The phrases "client-server" and "peer-to-peer" have become commonplace in network computing and usually describe some type of network function. However, client, servers, and peers actually describe roles played by participants in network communication. Because the roles of client, server, and peer can constantly change during a session, we can more accurately define clients, servers, and peers in terms of threads of execution that may exist on the same system or within the same process. A server opens a communication channel and waits for contact from a client. On the other hand, a peer may act as both a client and as a server.

Network Connections

Before we move to network architectures, it becomes useful to consider connections and the type of traffic found on networks. In this context, we can view the terms "peer connection" and "server connection" as defining the number of clients accessing a server on the network. Within this quick definition, a peer connection exists when a transaction occurs between only two computers located on the network. As an example, one network user may decide to share files located on his or her local hard disk drive with another user. Backing up data across a network between a server and a storage device located on the network also occurs through a peer connection. Because peer connections involve only two stations, the performance of the slower station limits the speed of the peer connection.

With the term "server connection," more than one client simultaneously communicates with a single computer operating as a server. With the server connection, only the performance of the server limits the capabilities of the connection. Here, the number of stations accessing the server negates the effect of any one client on system performance.

Middleware

As illustrated in figure 7.1, the term "middleware" defines a combination of an architecture, a programming language, a communications program, a data manipulation program, a programming interface, and a translation driver. This combination allows middleware to act as a bridge between systems and provides a method for allowing users to connect multiple

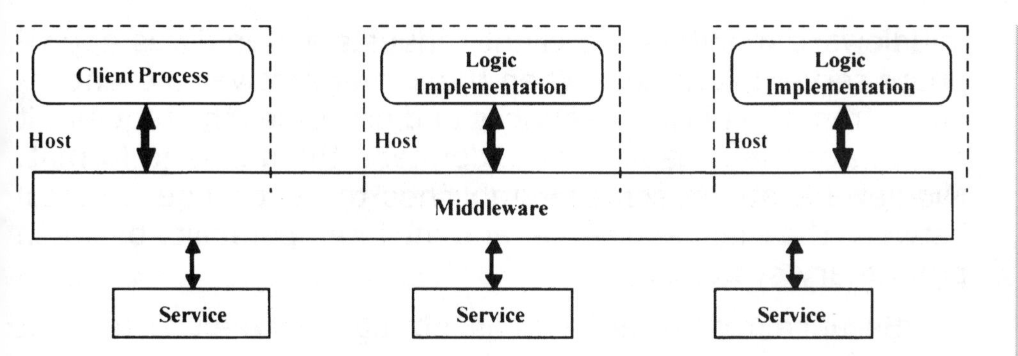

Figure 7.1. Middleware architecture

data sources through networks. Middleware smoothes the differences between communication protocols, database formats, operating systems, and applications while delivering the data for processing.

In addition, middleware establishes a feedback path to the operating system for the purpose of network management. The middleware also provides a variety of management and support services such as security services, directory services, and time services. Special-purpose middleware developed for emerging applications includes wireless middleware for mobile computing applications, middleware for distributed multimedia applications, middleware for groupware, and middleware legacy system access and integration.

Middleware enables applications and establishes sessions between client and server processes, security, compression/decompression, and failure handling. Operating through a set of software modules invoked by client processes and Application Programming Interfaces, client middleware provides the interfaces between client processes and remote server processes. For example, many database servers using the Structured Query Language use API/SQL software that a client process can utilize to send SQL statements to the SQL servers. Examples of client middleware include Web browsers, Open Database Connect, or ODBC, drivers, and the Common Object Request Broker, or CORBA, client middleware. Server middleware monitors the client requests and initiates appropriate server processes. When the server receives a client request from the network services and processes the request, it must have the capability to understand the message format. Middleware at the server establishes the exchange protocol between the client and server and the interoperability between proprietary systems.

Because middleware acts as a bridge between systems, it resides above the network and below the application software. Referring back to the OSI model, middeware operates between the Session Layer and the Presentation Layer. Most middleware expands and improves session connections while operating at the Session Layer. However, middleware also remains closely tied to Presentation Layer services and improves Presentation Layer services through the addition of format translation and presentation templates before data moves to the application. As a result, applications can have more network connections to data and code libraries.

Given this combination of tasks associated with middleware, three basic types of middleware exist. Database middleware establishes communication links with host databases, translates information obtained about the database type and structure, and translates the results into a user application format for additional programming or changes. The establishment of communication links with host databases involves the mapping of connection paths, the setting of links, the converting of protocol, and the delivering of data to local ports.

Messaging middleware supports interaction between messaging systems such as e-mail or scheduling software. As an example, electronic mail software provides business-unaware services that reside above networks and interconnect users or applications. Messaging middleware also handles interconnections between groupware products, Web browsers, Web gateways, SQL gateways, and Electronic Data Interchange packages. World Wide Web middleware includes the Hypertext

Transfer Protocol, the Hypertext Meta Language, the eXtended Meta Language, C++, and Java.

Distributed data and transaction management middleware operates with different types of data. In addition, distributed data middleware provides connectivity between document imaging and textbase systems, and allows the access, manipulation, and update of distributed as well as replicated data. Users may retrieve data dispersed around a network or across networks. As multimedia technologies continue to grow and gain popularity, middleware designed for multimedia services handles the movement of image databases between internetworked applications.

Peer-to-Peer Networks

Peer-to-Peer technology enables a direct exchange of services between computers. The services may include the exchange of information, processing cycles, cache storage, and disk storage for files. Within Peer-to-Peer, an organization can establish policies for the use of a navigation service to mediate or facilitate the direct exchange. With mediation, users register their own files and search for other files to copy. With the direct method, users register their files with network neighbors and search across the network to find files to copy.

Peer-to-Peer communication achieves distributed computing goals by allowing programs located on different computers to communicate through file transfer. On a large scale, Peer-to-Peer could allow some network traffic to move from the corporate backbone to a less expensive infrastructure consisting of switches, hubs, and routers. The client layer could offer large amounts of spare storage and unused processing power without placing any addition strain on the backbone.

In layered network architectures, a process may only communicate with a peer process. A Peer-to-Peer process pro-

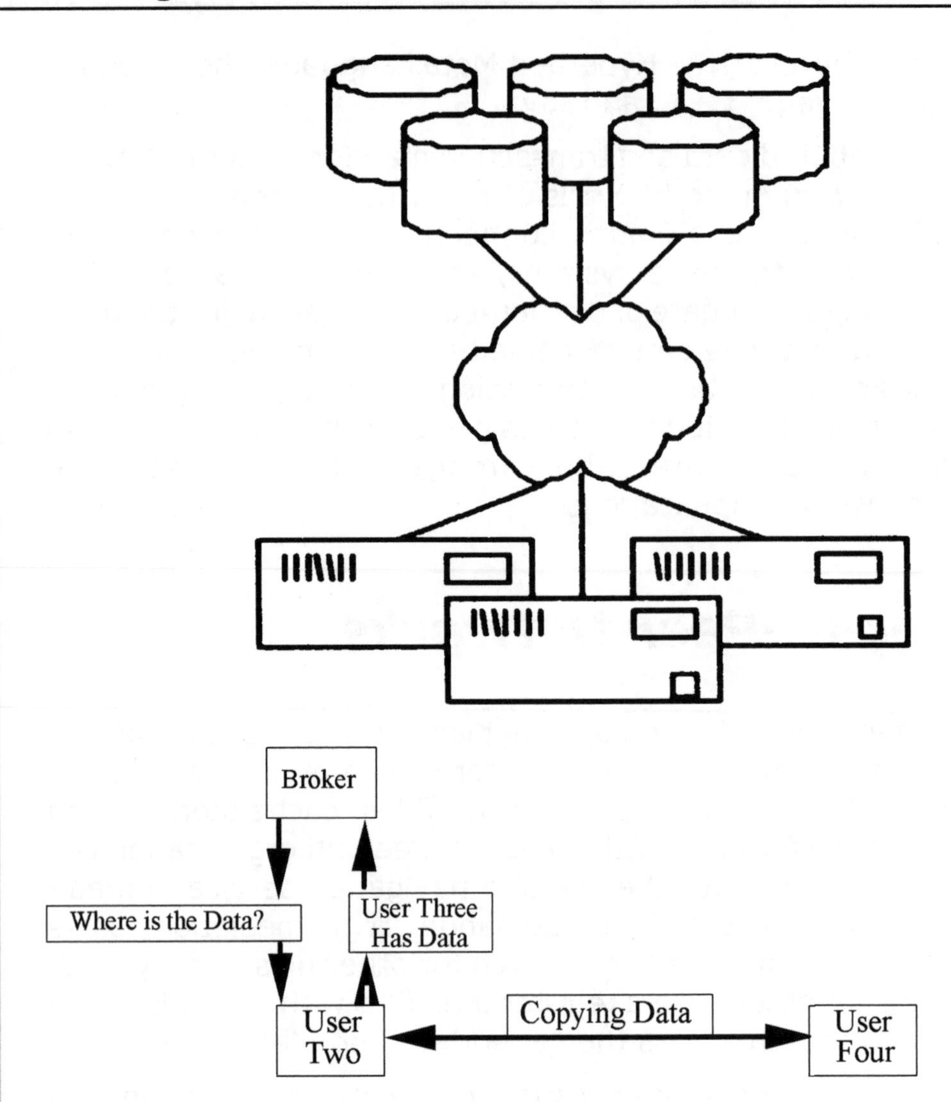

Figure 7.2. Peer-to-Peer networking in the enterprise

Figure 7.3. Peer-to-Peer broker-mediated file sharing

vides a communication format where neither process controls the other. In addition, the same protocol operates with data flowing in either direction. Peer layers communicate through a common protocol that matches the services provided by the layers.

The simple network shown in figure 7.2 functions as a Peer-to-Peer network. However, because of eCommerce and eBusiness, several variations of Peer-to-Peer networking have emerged. With the broker-mediated file-sharing model shown in figure 7.3, a set of servers and desktop computers operate together. The server provides database search capabilities and directs one peer to another. When a data request arrives at the broker from a client, the broker responds by finding the peer that contains the data and notifying the second peer about the first peer request. From there, the two peer computers exchange data.

Figure 7.4 shows an example of Peer-to-Peer file sharing. With this model, a requesting peer asks the next peer for data. The peers pass the data until finding a machine that contains the requested data. Then, a connection occurs between the requestor and the peer holding the data. Moving to figure 7.5, note that the cycle-sharing model relies on a master controlling server and the division of process queries into portions.

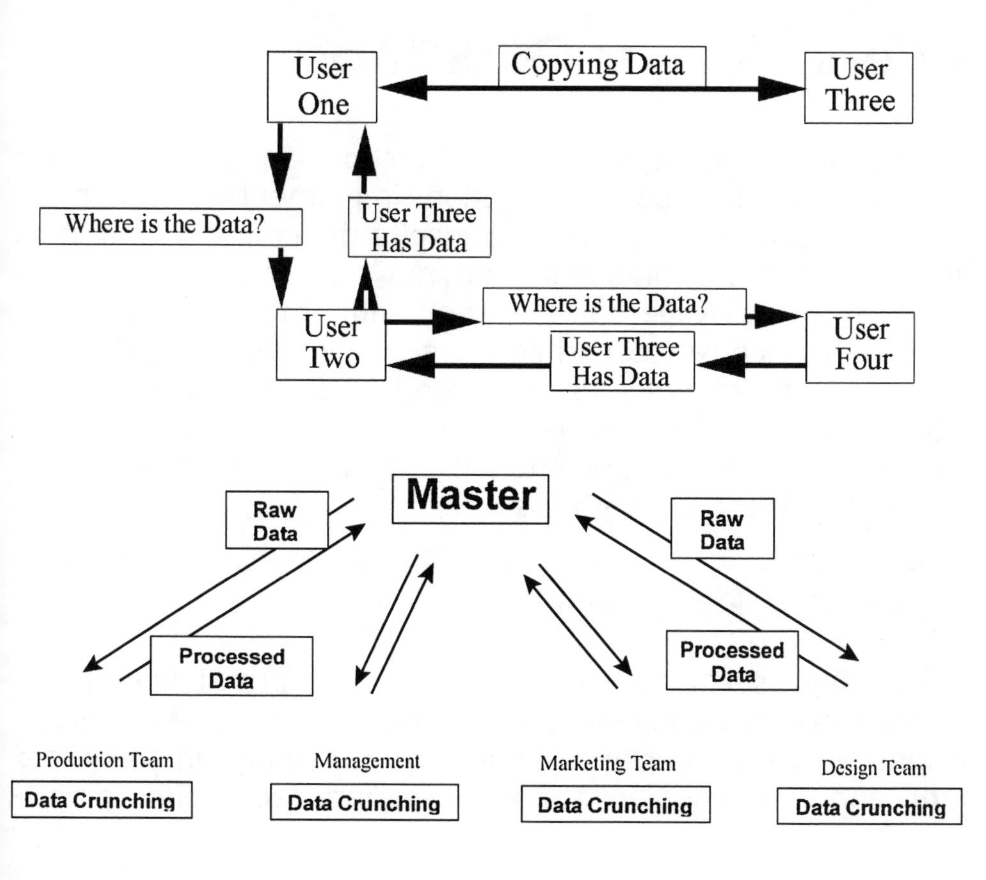

Figure 7.4. Peer-to-Peer file sharing

Figure 7.5. Peer-to-Peer cycle-sharing model

The master controlling server sends the process query portions to the clients. After the clients process the query and return individual query responses to the server, the master controlling server reassembles the individual responses into a total response.

With the cycle-sharing model, workstations on the network can access the computing resources of lesser-used equipment. As an example, teams working in the design stage of a project and requiring more computing resources can leverage the processing power of other teams who do not require the resources. During this process, a broker would take requests for computing resources, scan the network for workstations that have lighter loads, and send batch data to the machines for processing.

Client-Server Operation

Clients and servers operate as functional modules that have well-defined interfaces and functions implemented through the use of hardware and software. Despite the common definition for servers, clients and/or servers may run on dedicated computer systems. Regardless of the configuration, a client initiates a service request while the server responds to the service request. For a given service request, the equipment operating as clients and servers do not reverse roles. In contrast, though, a server can become a client when the original client issues a request to another server.

Client-Server Processes

A client-server process establishes a format where one process serves as a client while the other operates as a server. Client processes perform application functions on the client side and can range from basic user interfaces and spreadsheets to complete application systems. With the impact of the Internet, some client processes use Web browsers to interact with end users. In contrast, server processes provide a service to the client by performing application functions on the server side. Server process can range from basic functions to higher level applications such as order processing, electronic funds transfer, print services, database processing, object servers, and mail services.

Each client process uses a shared protocol to request that the server perform some task. Because the client-server process allows the sharing of networked resources, the client has the capability to take advantage of a resource — such as processing power, storage, or software — that resides on the server. Even though the processes share a common protocol, the protocol defines different methods to support communications sent from the client and communications sent from the server.

Client-Server Communication

Clients and servers communicate through a Remote Procedure Call, Remote Data Access, or Queued Message Processing. The Remote Procedure Call, or RPC, involves having the client process request a server process called a remotely located procedure. Each request or response within the RPC exists as a separate unit of work. As a result, each RPC must carry enough information to meet the needs of the server process. A remote procedure may occur as a basic or a complex task.

A Remote Data Access, or RDA, allows client programs and/or end-user tools to issue ad hoc queries of remotely located databases. Compared to the RPC, the Remote Data Access does not produce a result that has a known size. The database query could produce one row or thousands of rows within the response.

With the Queued Message Processing, or QMP, paradigm, the client message stores in a queue. When free of other tasks, the server processes the client message. From there, the server stores, or puts, the response in another queue and the client retrieves, or gets, the responses from the active queue. Used in many transaction processing systems, QMP allows clients to send requests to the server asynchronously and provides the benefit of processing requests even when the sender has disconnected.

Messages

The exchange of information between clients and servers occurs as messages. Each service request and the additional routing information becomes part of a message sent to the server. When the server responds to the service request from the client, the response becomes part of a message. In most instances, the client-server model supports interactive rather than off-line messaging. However, queuing systems allow clients to store messages for asynchronous pickup by servers at a later time. The use of interactive messaging increases the flexibility and scalability of client-server operations while increasing concerns about portability, interoperability, security, and performance.

Messaging allows an intermediate processing layer at the server to route the message to the appropriate receiver. Because message-based communication operates well with intermediate routing, the combined features can provide a higher level of abstraction for the communication framework. A server or router may deposit messages into a queue while one or more logical processors retrieve and act on the messages.

Depending on the type of operation, the processors may not respond to the messages at all or may respond directly to the client. To maintain the abstraction, however, the processors can send a message back to the server through another queue that routs the message back to the client. Message-based architectures may operate synchronously. In one type of synchronous mode, the server or router passes the message to a processor that passes a response back to the server. The server returns the response to the client. In another messaging format, the server operates asynchronously, and the client operates synchronously. Combining asynchronous and synchronous operation in this way allows the server to gain the efficiency of asynchronous operation while the client benefits from the procedural simplicity and safety given through synchronous processing.

Two-Tiered Client-Server Architectures

Initial implementations of client-server architecture used the two-tiered architectures shown in figure 7.6. In the figure, the two-tier architecture consists of a client operating at the front end of the network and maintaining application presentation and business application logic. A server operating as the back-end of the network provides data management functions such as data integrity, retrieving, adding, file removal, and updating. With the two-tier architecture, the client and server communicate directly with each other through a highly rigid and tightly coupled relationship.

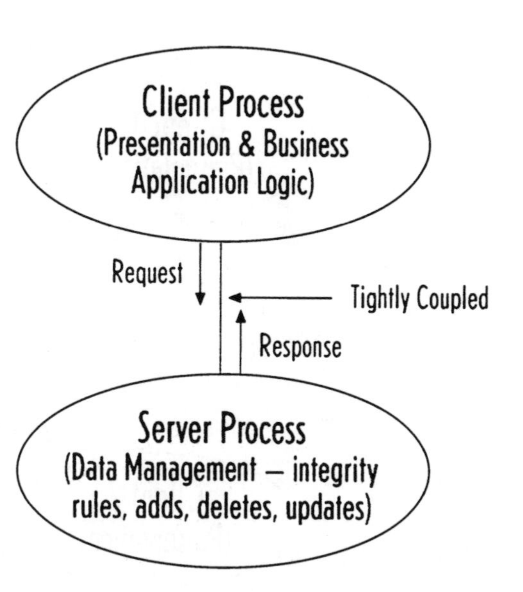

Figure 7.6. Two-tier client-server model

Additionally, the client and server must have fully synchronized communications. Because the client and server remain tightly coupled, the implementation of a change by one process requires the modification of the other process to accommodate the change. Because of this requirement, the two-tier architecture has higher maintenance costs.

Three-Tier Client-Server Architectures

As illustrated in figure 7.7, the three-tier client-server architecture increases the capabilities of the two-tier model by inserting a middle tier that maintains business logic, rules, and access to data from the server. With the insertion of this middle tier, the responsibilities for the client process reduce to presentation and user interface logic. A front- or back-end process implements through either a Dynamic Link Library or Application Program Interface. Access to the process occurs through Remote Procedure Calls or messages.

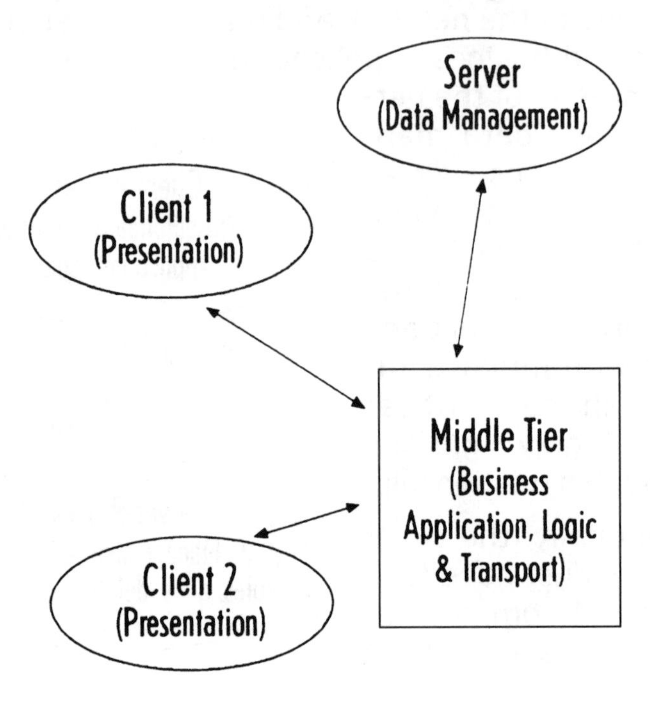

Figure 7.7. Three-tier client-server model

In all situations, the middle tier creates a more loosely coupled interface between the client and server tiers and between the business logic and the client Presentation and Transport Layers. Rather than requiring hardware support, the middle tier operates as a logical tier. With this, the middle tier process can run on either the client or the server. In some applications, the middle tier can compile into the client and server while providing a consistent and simultaneous interface to the server. As a result, many unrelated clients can use the same server. If a server users all the integrity rules for the serviced domain and all the client applications defer to the rules, all the applications will have consistent access to consistent information.

Layered Client-Server Architectures

Presentation, business application logic, and data management make up the functional portions of the two- and three-tier client-server architectures. With the layered client-server architecture, the functional portions separate into layers that provide specific functions within the overall application. The number of layers that occur within an application depend on the requirements of the application and the required amount of coupling between the layers.

In all but one instance, services spanning each layer provide either direct functional support to the overall application or a service to the other functional processing layers. The "respecification" layer offers the lone exception to functional support and provides the ability to adapt the front-end application without requiring significant changes to lower layers. As a result, the application can adapt to changing user requirements more quickly. Remaining portions of the architecture remain fixed.

When considering the services, the navigation service provides each functional layer with directions to the next layer. With this service in place, processing flows from one layer to another without fixed connections that could require reworking at a later date. Security services control the capability of the user to access various layers and execution services, but security also allows the verification of data access rights. Metering and logon/logoff services provide configuration control and an audit trail that assists with the fine-tuning of system performance.

Even though the number and type of functional support layers vary across projects, the basic design of the layered architectural model remains constant. As a result, the model can support the growth of an enterprise into a client-server environment. In addition, the use of the layered client-server model can establish the basis for adopting object-oriented technology.

Client-Server Software

The local software in an OCSI environment provides access and manipulation of data and processes located on the machines in a client-server environment. Examples of the local software include

- database managers,
- transaction managers,
- file managers, and
- print managers.

Database management systems, or DBMSs, provide access to databases for on-line and batch users. In a typical database environment, different users can view, access, and manipulate data contained in a database. With a DBMS, however, the system not only manages logical views of data but also allows different users to access and manipulate the data without knowledge about the physical representation of data. In addition, a DBMS manages concurrent access to data by multiple users. The security portion of a DBMS enforces logical isolation of transactions, limits access to authorized users, and provides integrity controls and backup/recovery of a database.

Transaction managers, or TMs, monitor the sequence of statements executed as a unit. In addition, the software manages transactions from the point of origin to the planned or unplanned termination. Some TM facilities integrate with the DBMS facilities to allow database queries from different transactions to access/update to one or several data items. In contrast, other transaction managers only handle transactions. With the application of transaction managers in a client-server environment, systems administrators face the challenge of managing the execution of transactions across multiple sites.

In a client-server environment, file managers allow the access and manipulation of text documents, diagrams, charts, images, and indexed files by multiple stations on a network. As the name indicates, different types of print managers control different types of printing operations. Almost all local-area networks provide access to print managers operating through the network.

Client-Server Networks

Conceptually, clients and servers may run on the same machine or on separate machines. In practice, however, clients and servers reside on separate machines connected through some type of network. With client-server networks, a group of personal computers attach to a server for the purpose of sharing resources. During operation, any of the personal computers — or clients — can request that the server perform a processing task.

With this, the task divides into two separate parts. As an example, a client process may create a report while the server process prints the report. The clients and server work through a shared protocol. Each transaction between the client and server takes the form of request/reply pairs where the client initiates a request and the server responds with an acknowledgment of the request. If the server cannot perform the request, it responds with an error message.

Operation of a Client-Server Network

In practice, a client computer has lesser capabilities but the server not only offers greater storage capacity, more random-access memory, and greater processing power, but may also use a different operating system such as Windows NT server or UNIX. Each client in the network includes a graphical user interface such as Windows or the Macintosh operating system that provides easy access to word processing, spreadsheet, e-mail, and presentation graphics applications while attached to the network. Each client on the network has simultaneous access to the applications found on the server.

Client-server networks may include different types of personal computers and often attach IBM-compatible and Macintosh computers to the same server. The software found at the server and client levels, the transmission media, and the network protocols establish cross-platform connectivity.

The client/server model describes communications between service consumers, or clients, and service providers, or servers. In addition, the model allows application components to operate as service consumers and service providers. Application processes between clients and servers represent the business logic as objects, may reside on different types of computing equipment, and may initialize through Web services.

Client-server applications employ this model to deliver business functions and provide a powerful, flexible mechanism for tailoring applications to business needs. As an example, a business could use the client-server model to maintain order-processing databases that contain customer and product records at the corporate office on a server. The same model would also implement and support order-processing logic and user interfaces for branch stores that initiate the orders on clients.

Client processes typically reside on desktop computers. In some applications, though, a client may reside on a mainframe. In this type of application, the client program may access a database server on a LAN. Examples of clients include

- Web browsers that allow Internet users to access information over the Internet through graphical user interfaces,
- decision support tools preprogrammed to access remote servers,
- purchased and/or developed client applications that access remote data when needed,
- object clients that issue messages to remotely located objects, and
- client application development tools that allow for the development of client applications on workstations.

When we combine the advantages given by the client-server architectures, two distinct advantages become apparent. Client-server networks allow for the distribution of applications based on requirements for resources. As an example, the server may provide processor resources, and the client may provide graphical display capabilities. Because a server process may serve many clients, a client-server network may also provide resource sharing.

In addition to function-specific activities, the computer providing the server processes combines with the network operating system to provide access management, backups, and redundancy. In brief, we can categorize server functions as

- local-area network servers that allow clients to share the printers and files,
- window servers that manage user screens on a workstation,
- Web servers that receive requests from Web clients to provide Internet services,
- name/directory servers that show the location of a named object such as a file or program,
- authentication servers that check the authorization of users to access particular resources,
- distributed file servers that provide transparent access to files allocated to different computers,
- database servers that take an SQL query and return the desired information,
- object servers that present an OO interface to the clients,
- transaction servers that receive a transaction and respond appropriately,
- application servers that provide a complete application in response to a request from a client, and
- groupware servers that manage text, image, mail, bulletin boards, and work flow.

As we consider client-server networks in terms of those processes, it may become easier to understand that client-server networks may include different types of personal computers. Many times, a client-server network will connect IBM-compatible and Macintosh computers to the same server.

The software found at the server and client levels, the transmission media, and the network protocols establish cross-platform connectivity. As an example, Macintosh may use the AppleTalk Filing Protocol, or AFP, whereas clients operating with UNIX may use the Network File System. Although manufacturers may develop client-server protocols for specific platforms, the marketplace has pushed vendor support of multiple client-server protocols. A traditional client-server vendor licenses, installs, and configures large enterprise software applications in organizations.

Clustered Architecture

A clustered architecture consists of a group of nodes connected by a fast interconnection network. With the flat clustered architecture, each node in the cluster includes an attached local disk array. As illustrated in figure 7.8, the nodes divide into logical front-end, or delivery, and back-end, or storage nodes. During operation, the front-end node receives data from the back-end nodes using a shared file system. In addition, each physical node can serve as either a front-end or a back-end node. All nodes in the flat architecture have identical properties and provide both delivery and storage functionality.

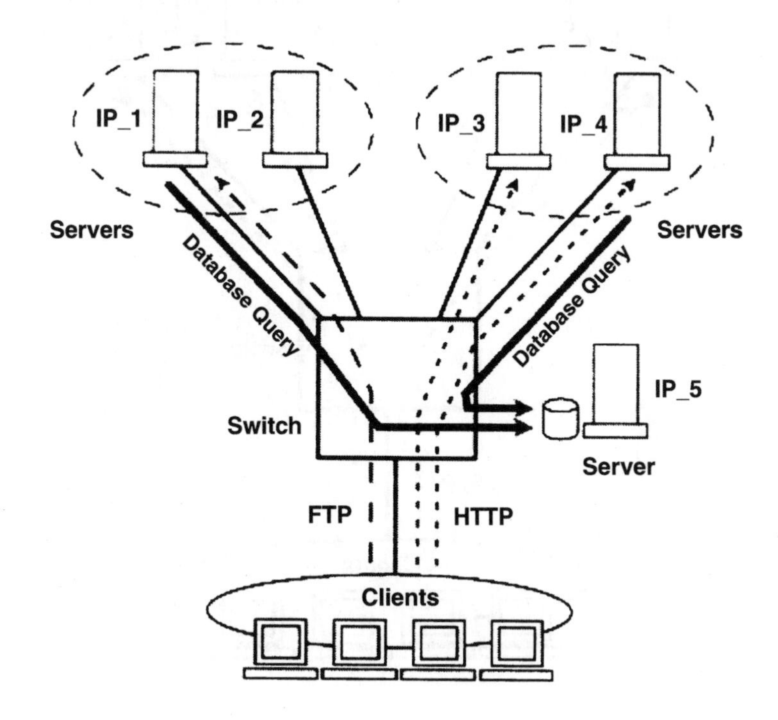

Figure 7.8. Cluster server model

Referring to figure 7.9, note that the two-tiered clustered architecture maps the logical front-end and back-end nodes to different physical nodes of the cluster. In this configuration, each node has distinct properties. The two-tiered architecture must include some type of underlying software layer — such as a virtual shared disk — that makes the interconnection architecture transparent to the nodes.

Increases in processor power have made server clusters a viable option for many organizations. A cluster consists of a collection of interconnected computers that operate as a single, unified computing resource. Each member of the cluster becomes referred to as a node. Shared resources in a cluster may include disk drives, network interface cards, TCP/IP addresses, and applications software.

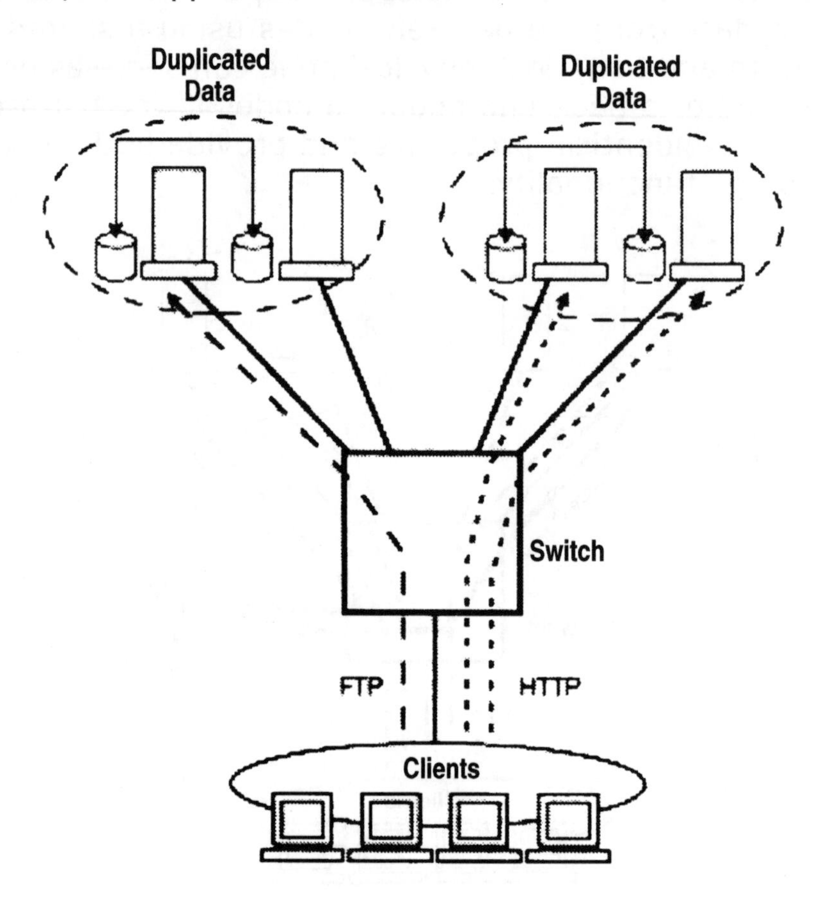

Figure 7.9. Two-tier cluster server model

Extensions to operating systems cause multiple computers to appear unified. The cluster service — a collection of software on each node — manages all cluster-specific activity while distributed file systems allow all elements of the cluster to gain access to data and programs. Clients connected to the cluster treat the group of computers as a single entity and can access files, resources, and objects residing at any location on the cluster.

Clustering technologies offer advantages in performance and availability. With a group of processors operating as a single unit, factors such as throughput and response time improve dramatically. Reducing the number of computers also eases cluster connection issues and provides high-speed interconnectivity. If one cluster node fails, the remainder of the cluster continues to operate and assumes the tasks performed by the failed system. In addition, clustered servers provide some detection and tolerance of software failures.

Defining Distributed Computing

The oldest form of distributed computing involves logging in to a host system from a dumb terminal or a terminal emulator running on a workstation. Although this method lacks the elegance of more modern distributed computing methods, the use of terminals and a host system involves a simple concept and basic implementation. The client operates as a directly connected terminal and has the capability to communicate through a remote connection. Pressing a key at the client sends a packet containing a code identifying the key to the server. In turn, the host system sends packets containing data back to the client for display.

Compared with the terminal interfaces, the Intranet, Extranet, and Internet technologies identified in the chapter introduction also use distributed computing technologies. A Distributed Computing System, or DCS, interconnects autonomous computers through a communication network for the purpose of supporting business functions. Rather than operate as a multiprocessor system, the computers connected in a DCS do not share main memory. As a result, information does not transfer between the interconnected computers through the use of global variables.

Instead, the exchange of information in a Distributed Computing System occurs through messages. With this setup, the computers must connect through some type of network. In addition, the interconnected computers must operate together and cooperate for the purpose of satisfying the needs of the enterprise. E-services built with agent technologies, traditional applications, and middleware components interact while offering services to consumers. In turn, a service stack consisting of host systems distributed across the network supports the applications. Figure 7.10 shows a representation of the service stack.

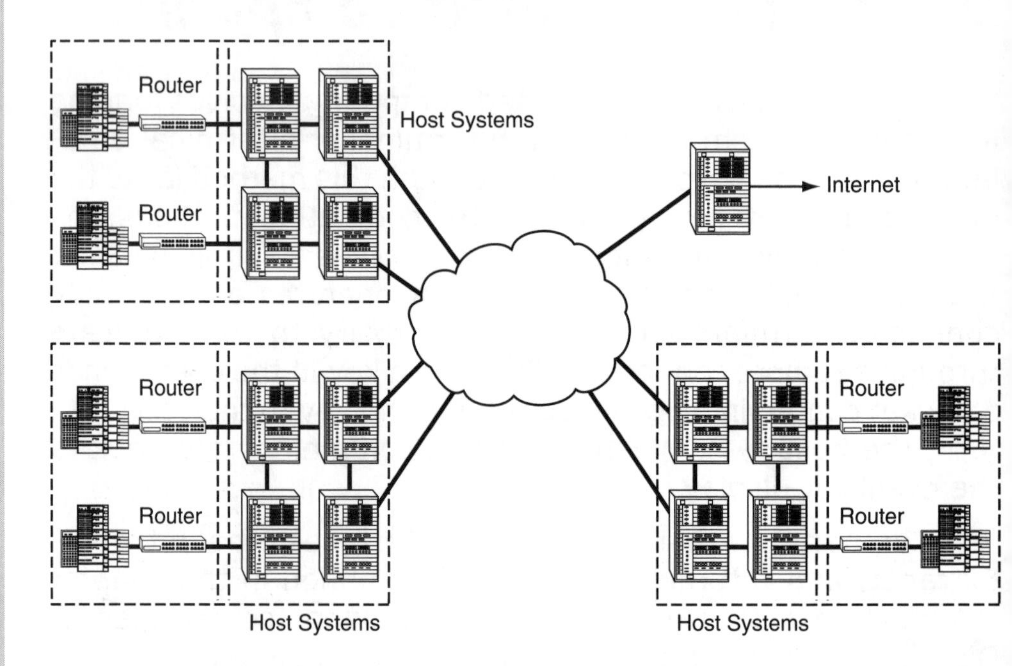

Figure 7.10. Host systems distributed across the network

With each of those resources, powerful and flexible tools based on Web architectures integrate legacy data needed by a small business. The use of these tools has prompted many small businesses to use hosted applications, or software applications and services delivered by an Application Service Provider, or ASP, that installs and hosts applications within a data center. Wide-area networking technologies employed through the Internet, an Intranet, or Extranet make the delivery of ASP applications and services possible.

Distributed Computing Environment

Most operating system vendors support the OSF 1 DCE, or distributed computing environment that formalizes many of the distributed computing concepts with a group of specifications. Although almost every available version of UNIX includes DCE core functionality, more advanced PC operating systems also include DCE core service support. Moreover, the DCE architecture defines thread, time, authentication and security, directory, and naming services.

Distributed Computing and Network Communication

The fundamental communication between networked computers at the transport layer of the OSI model establishes the foundation for all distributed computing architectures. Along with the communication defined within the Transport Layer, the communications protocol used for distributed computing depends on semantics, packet sequencing, data formatting, and other defined components. From there, the predefined protocols allow computer systems to interpret the packets received from other systems properly.

Messaging

As with client-server computing, distributed computing also depends on messaging or a process that labels a packet of data with the information it contains. Reliable messaging architectures appear similar to queued message frameworks. However, the messaging architectures have different modes of implementation. The reliable delivery of asynchronous messages requires the use of a "store-and-forward" model that synchronously passes the message to a middleware layer and then stores the message and any contained addressing information to a storage mechanism before returning control to the sending process. After the message has stored, the middleware can use different methods to send the message to its intended recipient. Throughout the process, the sender continues to process commands.

Store and Forward Messaging

The reliability of the store-and-forward model exists because the current holder of the message does not destroy its persistent copy of the message until it has received confirmation that a subsequent receiver has successfully stored the message. Because each link in the communication chain stores the message until receiving this confirmation, the original sender can proceed with its processing with the assurance that the message will reach the designated destination. The asynchronous nature of the store-and-forward model establishes a method where the sender must request confirmation of receipt of the message or receive confirmation when the message has reached the destination.

Distributed Computing Tools

Most distributed computing applications rely on the logic provided through an HTTP gateway application code written in Perl, TCL, or C. In brief, the gateway program connects to an application. Using a Web-enabled database as an example, the application builds Structured Query Language statements, executes the SQL statements, and formats the results. Other Web-enabled database applications rely on HTML pages that include module tags and functions that dynamically execute SQL statements specified by the user.

Hypertext Transfer Protocol and the HyperText Markup Language

The HyperText Transfer Protocol supports the transfer of hypermedia documents and forms the foundation for the World Wide Web, but the Hypertext Markup Language, or HTML, stands as the basis for publishing documents on the Web. HTTP consists of "send" requests from a Web browser to a Web server and the corresponding response. Rather than operate as a programming language, HTML is an application consisting of plain text embedded with tags that encapsulate the elements of a document. HTML tags consist of one or more characters and remain enclosed in angle brackets.

Extended Markup Language

Both HTML and the Extended Markup Language, or XML, evolved from the Standard Generalized Markup Language, or SGML. The phrase "markup language" refers to the simple notations used by an editor when preparing a document for publication. Because XML offers more flexibility, power, and complexity, it allows users to define individual formatting codes. In this way, every XML user can create usage rules.

Distributed Computing and Customer Needs

When considering new network technologies, a company that projects a stable growth rate and maintains a consistent workload may decide to implement the client-server application for several reasons. Software configured for client-server architectures will continue to provide viable solutions for traditional businesses with a centralized employee base, limited customer interaction, and predictable workflow. Customers have greater control over the hardware used in a client-server implementation than with a hosted solution. Because the client-server application and databases reside in the customer's data center, these systems support extensive customization. Finally, client-server applications do not require the Internet connectivity needed for the full functional operation of hosted applications.

On the other hand, companies that either must adapt to fast growth or highly variable workloads may not choose a client-server application that has limited future capacity. Organizations that operate with geographically separated locations or that rely on highly mobile workforces can make excellent use of hosted applications. In either situation, the utilization of distributed computing tools maintains consistent communication even when large-scale personal interaction does not exist. The consistent communication occurs through the full-time universal access to information regardless of location. The use of hosted applications promotes collaboration over the Internet because of the capability to share information securely with customers, clients, and vendors.

In addition to scalability, the installation of the ASP model has less complexity than the client-server model. Most ASPs use a standardized installation process that reduces the time between installation, configuration, and actual use. The preinstallation of a hosted application at the ASP data center allows new subscribers to use the application within in a shorter time. Moreover, many providers will assist new customers by loading the system with their pertinent data and configuring the application for the customer's needs. From that point, the customer can immediately begin training staff and using the application.

8

Object-Oriented Technologies and Application Hosting

Introduction

Internet, Intranet, and Extranet solutions have become prominent methods for quickly providing information to clients, potential customers, or employees. Technology tools that support Internet applications have given companies access to images, sound, formatted downloadable documents, and spatial data. Specifically considering Intranet solutions, a company can distribute project management information or other information such as employee directories, human resource materials, field critical information, or marketing plans to employees working in local locations. The use of an Extranet allows customers, partners, and suppliers to purchase products and services, learn about new developments in the business environment, and interact with on-line forms and other documents.

Object-Oriented Technology

Distributed computing applications take advantage of the flexibility given through objects, or reusable software components, and object-oriented programming. As illustrated in figure 8.1, each object consists of a discrete module that performs as a unit. Object-oriented programming uses preexisting objects and a building block approach to create programs. Programmers identify an object based on its input, output, and internal code.

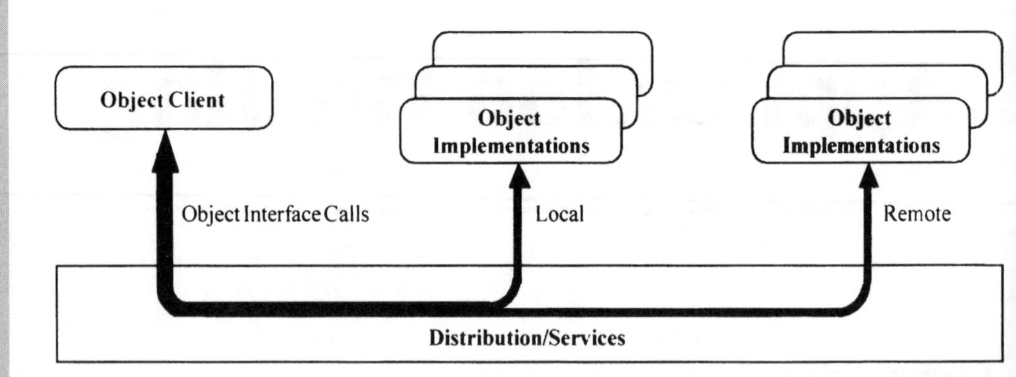

Figure 8.1. Distributed object architecture

The design of an object-oriented system depends on the ability of the developer to identify the objects in a problem, specify the attributes of the objects, specify behaviors associated with the object, and specify interaction between objects. Using that information, the programmer creates object classes.

Object Classes

With each object belonging to a class, the design gains a layer of organization. We can define a class as any uniquely identified model of a set of logically related instances that share the same or similar characteristics. As figure 8.2 shows, classes

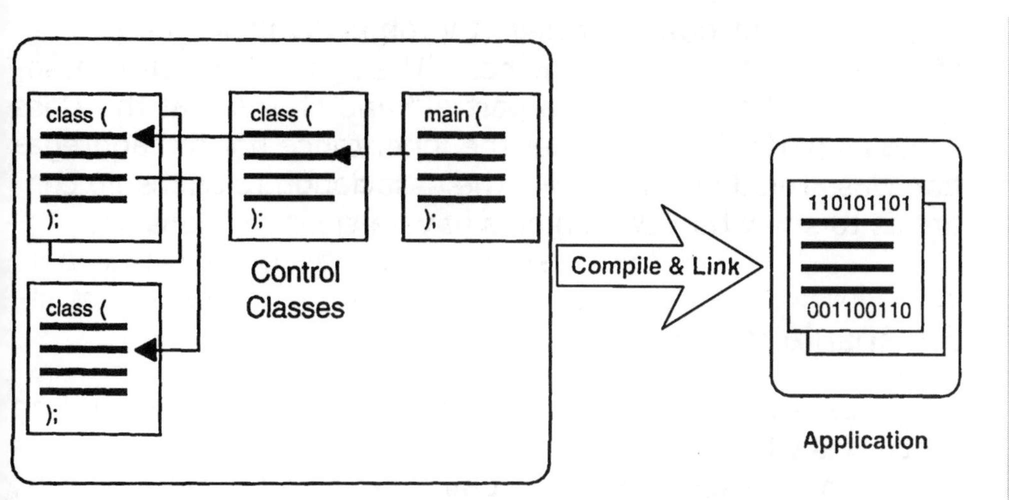

Figure 8.2. Use of object-oriented control classes to create an application

have attributes, methods, and a definition that allow the building of an application. As an example, an object class named *Student* might have attributes such as name and address and methods including *Add*, *Update*, and *Delete*. The definition defines the class attributes and methods.

Inheritance, Aggregation, and Association

Inheritance links an object with the phrase "is a kind of," and creates a flow chart of classes according to groups. As an example, the *Student* class would extend from the *Community Member* class. Given the characteristics associated with inheritance, programmers or users can group related items together and apply common definitions to the groups. In addition, the groups can accept individual variations of objects within the same class.

Inheritance allows the reuse of data definitions and program logic. Because of this property, developers can reduce the level of programming effort required to complete an application. Rather than redefining object characteristics shared between different object classes, the objects inherit the characteristics from one level to the next. The aggregation class associates with the phrase "is a part of" and shows that the class contains parts. When neither the inheritance nor the aggregation class relationships apply, the association relationship continues to show that two objects have a clear relationship.

Encapsulation

Encapsulation isolates programmers from the operating processes within the object and essentially labels the contents and the processes of the object as "unknown." Because of this, encapsulation shows that a specified action on the object produces a desired behavior. Programmers can assign methods to produce a certain object behavior; however, the programmer does not have specific knowledge about the logic that causes the behavior. Therefore, the object encapsulates its internal behavior.

Encapsulation protects the implementation of an object behavior against any unintended actions or inadvertent access. Because each object remains self-contained, a problem in one program area does not affect other areas of the program. In addition, a change in one element of an application does not cause problems in other portions of the application. From the programmer's perspective, encapsulation allows programmers to focus on larger scale issues and ensures that the code remains bug-free.

Polymorphism

Polymorphism shows that — depending on the situation — the same term can refer to multiple processes. With the multiple use of terms, polymorphism brings the concept of natural language to objects. Rather than rely on a set of sequential steps as seen in traditional programming applications, polymorphism hides the logic for performing the sequential steps within the execution engine. Even though the hidden code defines specific steps for a particular operation, the external operation of the object has a much simpler appearance.

Communication Between Objects

Objects communicate with one another through message passing. Communication usually consists of one object requesting some type of action from another object. Message transmissions between objects may take either a synchronous or an asynchronous form. With the synchronous form, the object enters a wait state until receiving a response. With the asynchronous form, the object continues to execute subsequent instructions.

Component Broker Technologies

An object-oriented program can execute only if activated within a compatible runtime environment. Called a component broker, the runtime environment operates as object-oriented middleware. Component brokers provide intelligence so that objects can interact with one another while performing a task. The runtime environment covers everything running within the environment and facilitates functions that cross application or operating system boundaries.

Component brokers operate at the distributed enterprise level and contain multiple types of objects. Connectors found within the component brokers allow non-objected-oriented components of an information resource to cooperate with the object-oriented environment. Component brokers also support Object Request Brokers, or a special type of application object that handles the translation and transportation of requests and responses between clients and servers.

Figure 8.3. RMI distributed application

As a response to the limitations of the two-tier client-server architecture, component broker technology has become a focal point for the deployment of Web-enabled applications. Component broker technologies can expand Web applications to provide client-server applications. In turn, the expansion of Web-delivered applications sets the foundation for distributed computing and Application Service Providers.

Java RMI

Java has gained a great deal of acceptance due to its platform neutrality, safety, and object-oriented design. Because Java operates as a complete execution environment rather than a programming language, Java can provide a consistent and abstract interface regardless of the underlying platform. This platform independence occurs through the use of a Java Virtual Machine, or JVM, that emulates a computing platform. Each combination of hardware and operating system software that runs Java matches with a JVM. Because Java programs appear — at the application level — to run on the same computing platform, communication between Java applications becomes easier.

As illustrated in figure 8.3, Java RMI provides a language-specific architecture that eases the building of Java-to-Java distributed applications. The use of Java RMI when designing a pure Java distributed system provides the advantage of applying the Java object model. However, the reliance on Java precludes using Java RMI in multilingual environments. Yet, the platform independence associated with Java continues to allow deployment in heterogeneous environments.

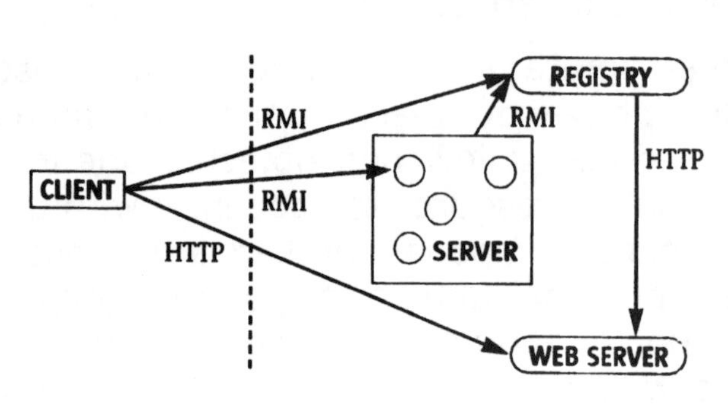

Microsoft Component Object Model and Distributed Component Object Model

Microsoft's Component Object Model establishes a software architecture that allows for the building of applications from binary software components and forms the foundation for higher-level software services. Rather than creating an application, COM services promote interaction between preexisting application components written with conventional development tools. As a result, COM allows a user to combine independently designed software development tools within the same address space. Using the same methods, the Distributed Component Object Model, or DCOM, works as an extension of COM and permits interaction between objects that execute on separate hosts in a network. With this approach, DCOM distributes objects across multiple physical address spaces.

Microsoft Object Linking and Embedding

Microsoft's Object Linking and Embedding Database, or OLE DB, provides a COM building block that stores and retrieves files to ensure connectivity. OLE DB provides access to corporate data regardless of the format or location and features full integration with the Open Database Connectivity structured query language interface and Microsoft's application development products. In addition, OLE DB simplifies programming, utilizes components that act as virtual tables for application development, and features interoperable data-centered components.

Middleware Tools and Technologies

Typically operating through three communications modules, a presentation component, and an application component, middleware includes five major technologies and tools that establish interconnections between applications and the redeployment of system processing. The communications modules direct the interconnection process through the use of transport components. While the presentation component delivers data into the client environment in a format compatible with programming applications, the application component providers additional structuring of the logic and functionality of the data.

Every middleware product uses the technologies and tools found within the three modules. The technologies and tools interconnect data and processing resources across distributed networks. Although the following descriptions provide separate definitions of the tools and technologies, most overlap or serve as a foundation for the remainder of the technologies and tools.

Applications Programming Interfaces

An Applications Programming Interface provides an interconnection based on programming between two applications. As the usefulness of APIs has grown, operating systems, developer tools, applications, and application drivers have integrated the middleware technology. APIs specify low-level connections between program components, integrate low-level components into a higher level, and create transparent interfaces and cross-platform programming capabilities.

Both the operating system and a set of network requestor APIs establish core communications within distributed computing environments. As illustrated in figure 8.4, an API contains groups of functions that transmit and receive bytes of data between systems when called by a program. Operating as low-level components of the overall system, the APIs provide the lower layers of the communication session. These low-level components provide limited abstraction of the underlying communication session and allow the programs to provide all logical services such as addressing and data conversion.

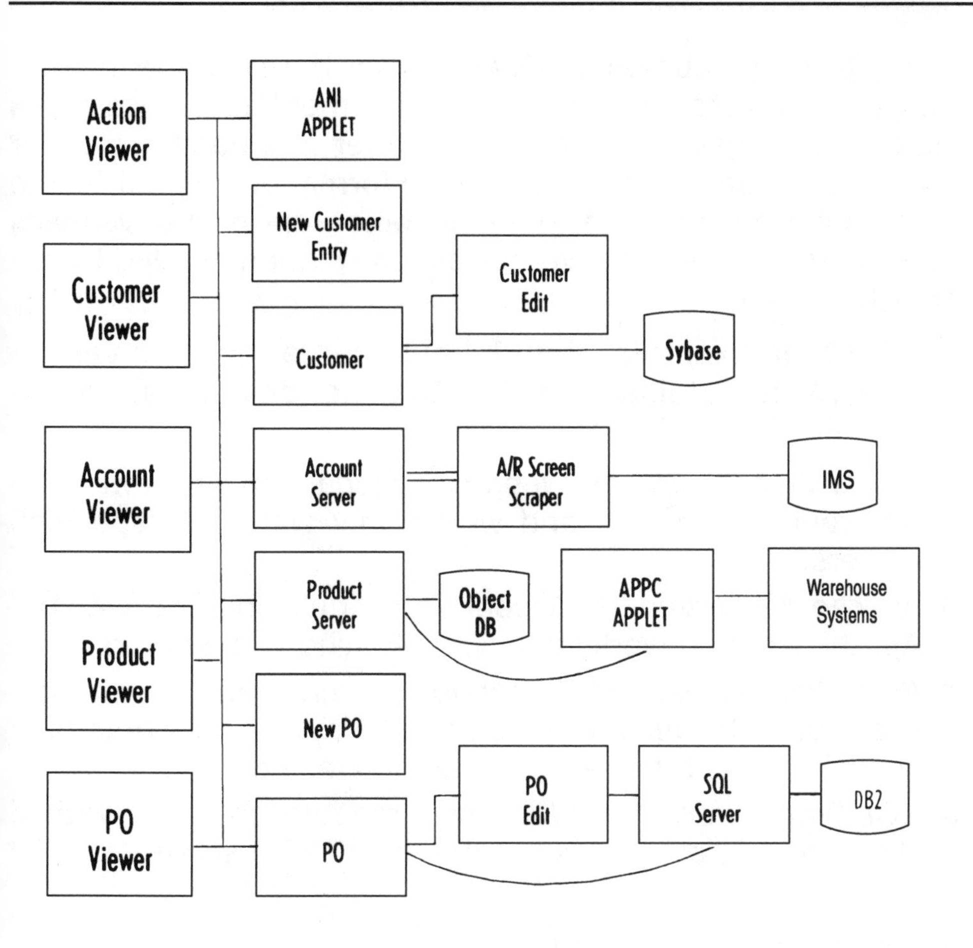

Figure 8.4. Customer API

The application illustrated in figure 8.4 includes five viewers that provide an interface for human interaction. When a user logs into the application, the action viewer and any applications available to the user initiate. Each application sends a message to the action viewer along with icons to display and a message that the user will receive when activating the icon. When the system receives a telephone call, the Automatic Number Identification, or ANI, applet sends a customer inquiry message to the customer domain server. The server accesses the customer database, packages the customer message, and sends the packaged message to the Customer Viewer. As with the action viewer, the ANI and the server initiate at the user logon.

In turn, the Customer Viewer parses the message for keywords to populate the screen. The Customer Viewer also sends inquiry messages to the Account Server and Purchase Order for the purpose of displaying any information related to the customer. When the company's customer representative fields the call, the representative has all information needed to answer the caller.

Another example of API implementation involves an Informix Web Datablade module that includes APIs for interaction with four different programs.

- *NSAPI Webdriver*, The Webdriver module operates with the Netscape Server API and works only with Netscape Web servers.
- *Apache Webdriver*. The Webdriver module operates with the Apache API and works only with Apache Web servers.
- *ISAPI Webdriver*. The Webdriver module operates with the Microsoft Internet Information Server API and works only with Microsoft Internet Information Web servers.
- *CGI Webdriver*. The Webdriver module operates with a standard CGI program executed by all Web servers.

Remote Procedure Calls

As an example of the impact of the DCE specifications, the Remote Procedure Call has gained industry-wide usage across heterogeneous networks. The concept behind the RPC takes an apparent normal procedure call from within a process and executes the call within some other process that may occur on a remote system. RPC protocols have the purpose of reducing communication complexity between processes through implementation hiding.

To accomplish this purpose, RPC mechanisms serialize function call data into a sequential stream and then reconstruct the data on the receiving end of the connection. Because the RPC operation occurs synchronously, it mimics the semantics of traditional procedural programming. The RPC client process makes a call to an apparent standard function known as a stub. Rather than execute the calls locally, the process passes the parameters to the function.

From there, the function packages and transmits the parameters to a remote execution environment. Then, the parameters pass to the real implementation of the function. After completing the execution of the function, the serialized return value passes back to the client stub. At this point, the RPC responds to the caller.

Interprocess Communication Tools

Interprocess Communication, or IPC, tools use multiple active tasks to synchronize processes dynamically and to pass information and messages between operating systems. In addition, IPC tools establish shared memory areas and processing space. As a result, the use of Interprocess Communication tools has become the foundation for all middleware applications and the impetus for distributed computing.

Message-Based Middleware

Message-based middleware distributes data messages containing the actual data requested and transmitted from sources. In addition, the middleware technology controls the security and reliability of interconnections and ensures that the connection exists at all times. The control messages contain addressing or processing information and provide the information to the recipient program. With this capability, message-based middleware establishes an environment that allows developers to connect the data and code from one system to another system. This connection occurs by using defined programming and processing interfaces found in the middleware.

Object-Brokering Tools

Object-brokering tools extend the capabilities given through remote procedure calls. The special-purpose middleware supports distributed object applications by allowing remotely located objects to communicate with one another. Examples of these tools include the Object Management Group's CORBA and Microsoft's ActiveX. Both of these middleware packages use the distributed object model based on the Object Request Broker, or ORB, that receives an object invocation and delivers the message to an appropriate remote object.

CORBA

The Common Object Request Broker infrastructure allows applications written in different languages to communicate and collaborate with each other regardless of the location. To accomplish this task, the application set to use the services provided by another application relies on the unique application code name called the Interoperable Object Reference, or IOR. Server applications can launch on demand. In addition, an application can simultaneously act as a client or a server.

Maintained as a standard by the Object Management Group, or OMG, CORBA allows the distribution of objects across heterogeneous networks. The object-brokering tool routes and manages objects and allows transaction processing and network management. Because CORBA offers a platform-neutral infrastructure for interobject communication, it has gained widespread acceptance. Applications use a common interface defined through the CORBA Interface Definition Language, or IDL, to communicate across multiple platforms and through various development tools.

Even though the IDL remains platform- and language-neutral, the Object Request Broker handles any data and call format conversions. In addition, the IDL specifies all interfaces to CORBA objects along with the data types used in those interfaces. Because of this common definition, the implementation and location of the CORBA object remains opaque to a client application. The CORBA Naming Service provides the method for locating objects by name. With any CORBA object, the client only has knowledge about the public interface and functionality of the object.

The Interface Repository, or IR, and the Dynamic Invocation Interface, or DII, also provide methods for identifying the capabilities of a runtime object interface and method of invocation. To allow dynamic global interaction, references to objects can pass among objects. As an example, an object referenced to a second object can pass the reference to a third object. During operation, CORBA must allow the third object to use the received reference to invoke the second object.

ActiveX

Introduced during 1996, ActiveX has become Microsoft's key distributed object and World Wide Web strategy. ActiveX provides a complete environment for components and distributed objects. All ActiveX components communicate with each other by using Distributed Computing Object Management, or DCOM — the Object Request Broker of the ActiveX environment. Consequently, Web technologies such as browsers, HTML pages, and Java applets can combine with desktop tools such as word processors and spreadsheets for distributed applications.

With this in mind, the Microsoft Internet Explorer Web browser can contain components such as Word documents, Java applets, C++ code, and Excel spreadsheets. During operation, a Java applet can call a remotely located Microsoft Word document through DCOM. Moreover, ActiveX clients can invoke the services from SQL servers and legacy access gateways. Each layer contains more functionality and eases the development and deployment of distributed applications. APIs provide access to the services provided through each layer and made available to applications, and network programming services invoke network services through APIs.

Object-Oriented Distributed Computing

Object technology is a key to the implementation of enterprise-wide distributed computing. Object orientation provides an approach to application development and database management systems. Seen as a method for simplifying the application programming process, object-oriented application development eliminates the need to consider procedural details of an application. With this, object-oriented application development provides an environment for building a complete application through the assembly of a collection of prebuilt parts.

An object-oriented database management system, or ODBMS, provides a method for storing objects. The objects may include traditional types of data stored in relational databases or different data types such as graphics, video, documents, or user-specified data. The design of the ODBMS stores the object as well as the methods for accessing, retrieving, displaying, and manipulating the object. Universal database management systems — such as those used on Web servers to store and manage Web-related data — integrate object-oriented capabilities with relational DBMS capabilities.

Distributed Objects and Application Hosting

We can define distributed applications as a collection of objects that incorporate user interfaces, databases, and application modules. Each object includes individualized attributes and uses methods that define the behavior of the objects. For example, an order found within a distributed object may involve the data and the methods that create, delete, and update the order object. The modeling of interactions between the components of an application can occur through "messages" that invoke appropriate methods. More specifically, the modeling of distributed applications may include class and inheritance concepts that lead to the type of reuse and encapsulation critical to managing the complexity of distributed systems. Table 8.1 illustrates some rules of distributed applications.

A customer can be defined as a class from which other business classes that define different types of customers can inherit properties.
An inventory can be defined as a class from which other properties of specific inventory items can be inherited.
An entire legacy application can be viewed as an object (or a class) by using object wrappers that mediate between legacy systems and object-oriented users.
Objects are data surrounded by code with properties such as inheritance, polymorphism, and encapsulation. Objects can be clients, servers, or both.
Object brokers allow objects to find each other in a distributed environment and interact with each other over a network dynamically. Object brokers are the backbone of distributed object-oriented systems.
Object services allow the users to create, name, move, copy, store, delete, restore, and manage objects.

TABLE 8.1 — DISTRIBUTED APPLICATION RULES

Object distribution architectures build upon the middleware concept by encapsulating data within functional interfaces to objects. Operating much as an API, the distributed object includes implementation details that remain hidden from the user of the object. However, the comparison ends because object architectures limit access to the invocation of methods defined for the object. In addition, the invocation of an object occurs indirectly through references to the objects and eliminates the need for local instances of the objects. As a result, distributed object architectures can support location, platform, and programming language transparency.

One challenge that exists with enterprise-wide distributed applications occurs with the extension of object-oriented concepts to the distributed computing environment. Distributed objects disperse across the network and allow access by connected users or applications. As such, enterprise-wide applications can exist as objects residing around the network.

During operation, an object on one machine can send messages to objects residing on other machines. As a result, the object sees the entire network as a collection of objects. With this, the concept of distributed objects evolves into object frameworks, business objects, and component software for distributed systems. The use of distributed objects addresses problems such as reuse, portability, and interoperability because applications constructed from reusable components that include internal details can interoperate across multiple networks and platforms.

Application Service Providers

Application Service Providers offer the services through a monthly lease schedule that offers cost savings when compared to traditional ownership, installation, and support costs. Application servers not only provide a framework for rapid deployment of new business processes but also separate business logic from back-end data. This capability makes it possible to outsource the application server and the processing of the middle tier business logic to an application service provider.

In many ways, Application Service Providers represent the first moves toward distributed computing and away from the common client-server environment. Because of the services seen with an ASP, distributed computing offers the benefits seen with centralized computing without sacrificing flexibility or control over local computing resources. In effect, the business gains freedom from the administration and management of key enterprise applications while retaining microcomputing resources and maintaining independence from traditional centralized computing.

Distributed computing combines the advantages given by an existing Internet infrastructure and extends that advantage through existing corporate connections to that infrastructure. Because the use of Web-hosted applications requires no additional server, client, or network investment, an organization can decrease the investment into maintaining, accessing, and utilizing information technologies. In comparison to client-server applications, distributed computing does not require the use of IP addresses or port and socket configurations for remote access.

The universal qualities of the Internet translate into geographically distributed, controlled access to business information. Colleagues, contractors, and customers can access this information from any location that has access to the Internet. Because the ASP data center stores user preferences and data, the appearance and usefulness of the system remains the same regardless of the location. In addition, distributed computing maintains the capability to share information and collaborate on a project from any location. Because Application Service Providers establish connectivity through standard Web protocols, firewalls do not restrict access.

An Application Service Provider will want to offer its hosting service to many clients, and keeping the application components and data belonging to different clients separate will be essential. All this requires fine-grained access control in front of the application server. In a service provider environment, using separate machines for different clients becomes a burden because of physical space requirements.

As a result, some ASPs turn to a solution that involves the use of fewer, more powerful machines. However, hosting different clients on the same machine requires special security precautions for separation. To accomplish this separation, an ASP may press a trusted operating system that can handle classified data into service. The use of this type of operating system enables the configuration of the network for the required separation.

An Application Service Provider hosts applications that integrate processed results into applications running on a server located at a business or organization. Each application or service involves the leasing of network-based software tools as a replacement for processing capabilities typically purchased, licensed, or developed by a company. As a result, application services such as human resource or accounting tools have gained attention from small and midsized organizations and businesses because the ASP maintains key technologies and applications. Using Web-enabled technologies for the outsourcing and hosting of applications and services provides different levels of flexibility and interaction.

Application Outsourcing

Application outsourcing delivers enterprise applications from the provider to the business on a lease basis. On a broader level, application outsourcing can combine with the outsourcing of data networking services, hardware and software maintenance, systems administration and maintenance, messaging, and other information technology services. In each instance, the provider also offers the skills and capabilities to maintain mission-critical systems.

Application Hosting

Web technologies have enabled interactivity and transaction-handling capabilities that can integrate with existing local systems. Web hosting involves offering a suite of on-line applications for a specific group. As an example, an on-line broker could offer applications such as statistical analysis, instant information, and graphical representations of trends.

Another solution involves hybrid ASP-enabling technologies. As an example, a commerce hosting service may offer not only the platform for building the application but also credit card processing, data encryption services, specific interfaces designed for enhanced customer service, and other specialized transaction services.

Web Hosting

Web hosting involves the maintenance and operation of a Web server by a Web Presence Provider, or WPP. In addition, Web hosting designs, integrates, operates, and maintains all the infrastructure components required for running Web-based applications. As an infrastructure service, Web hosting covers network access, data staging tools, and security firewalls.

Typically, a Web Presence Provider has one or more data center facilities connected to the Internet. However, the definition of a Web Presence Provider may vary from a simple facility where a T1 line connects to the server and an Internet Service Provider. In other instances, the WPP may offer high-speed Internet connections and redundancy protection against disaster.

Web hosting operates through the use of virtual servers that create an illusion of more Web sites than actual servers. To create this illusion, the Web hosting service places a set of virtual servers on the same server by manipulating IP addresses. With one method, multiple domain names assign to a single IP address. Although multiple logical hosts share one physical host, the assignment of multiple domain names to a single address provides the appearance of each host having a Web server. Grouping several virtual servers on the same server also allows better reuse of extra server capacity.

A second method creates a separate IP address per each host. The use of a separate address per host allows the system to overcome limitations on the number of hosted sites. However, the method reduces scalability when operating with a large number of small sites. A second method creates a separate IP address for each host. The use of a separate address per host allows the system to overcome limitations on the number of hosted sites. However, the method reduces scalability when operating with a large number of small sites.

In a shared hosting environment, one computer system hosts many different Web sites. Given the nature of the shared hosting environment, both the provider and the customer achieve economic benefits. The total cost of hardware, software, maintenance, operation, and customer support divides among customers. Configuration can involve the use of a single configuration file that contains all Web site information and no limits on the use of system resources. In most instances, however, configuration involves the use of a single configuration file for Web sites not requiring many system resources and multiple Web servers for other resource-consuming Web sites.

In the dedicated hosting environment, one computer system hosts one Web site. As a result, the environment offers a combination of flexibility and security for the provider and the customer. In addition, the dedicated hosting environment standardizes the software configuration and dedicates all system resources to one Web site. A dedicated hosting environment includes

- hardware,
- an operating system that incorporates TCP/IP,
- an Internet connection consisting of an IP number and domain name, and
- Web server software that uses protocols such as HTTP and FTP.

To handle a large number of clients concurrently, Web servers operate with either a multithreaded or a multiprocess model. Multithreaded Web servers require the kernel thread support provided through operating systems such as Solaris and Windows. During operation, the server assigns a separate kernel thread to each incoming HTTP request. Threads can share a common state in global memory.

In the multiprocess model seen most often with UNIX applications, a separate process assigns to each incoming request. Because the initiation of a process demands resources, the startup of the server creates a pool of processes. Created processes operate on a common Web server socket and may communicate through the use of shared memory. A process that accepts a connection handles it until the connection closes.

Server Farms

As the demand for access to Web-delivered information has increased, Web hosting has evolved from a reliance on a single computer to the use of redundant, load-balanced server farms that create scalable solutions and a clustered architecture that consists of a group of nodes interconnected into a high-speed network. With either infrastructure solution, each Web server can access the entire content of the World Wide Web and can satisfy any client request. The distribution of traffic to and from a Web site across multiple servers located in different geographic locations allows the addition of system resources without interrupting the Web-hosting service. In addition, the Web site remains available and accessible even when LAN or Internet connection problems occur.

Application Services Benefits and Costs

A business or organization that builds additional technology outsourcing around distributed computing can emphasize business processes and workflow rather than splitting resources between those needs and technology needs. As already mentioned, the emphasis on hosted applications may reduce long-term costs associated with license fees, the outright purchase of server and client computer systems, and system administration. The ASP spreads those costs across a base of multiple users.

In addition, the reliance on ASP-delivered services and applications moves the responsibility for installation, maintenance, security, and scalability to the vendor and away from internal departments of the company or organization. Each of the components derived from software components found on the traditional client-server network can remain accessible through applications hosted and maintained by the ASP data center. These applications include relational databases, application servers, and Web servers. In addition, the ASP can provide system management software that protects critical data through backup and security utilities.

However, an emphasis on distributed computing also increases a company's or an organization's reliance on third-party resources and the access given through the Internet. In some cases, a company may also opt to switch from sole reliance on Internet access to access through a private network based on leased lines that gives access to the needed resources. When the Internet provides the capability and the quality-of-service guarantees, the Internet can exist as the communications infrastructure for mission-critical ASP services. In contrast, many ASP companies provide maximum responsiveness to an application and quality-of-service by either establishing leased lines or colocating servers within close proximity to the customer.

When comparing the software upgrade and on-line documentation implementation timeframes and ease with traditional client-server operations, the hosted application offers several advantages. The installation of a hosted enterprise application occurs within the ASP data center. As a result, the ASP also upgrades the software when necessary and changes on-line documentation to match any upgrade or solution needs.

When upgrading software, an ASP data center uses a consistent, standardized content-staging system for developing, testing, and deploying new modules. After completing the upgrade, the vendor notifies users about the upgrade completion and the functions contained within the upgrade and provides training materials that illustrate the enhancements.

The client-server application upgrade process requires more time, more resources, and more user acceptance than a hosted application requires. To upgrade a client-server application, the vendor must perform rigorous cross-platform testing, burn new copies of the installation media and documentation, and assist with the installation and configuration of the upgrade. In addition to supporting shorter development cycles, a hosted application supports seamless integration with best-of-breed third-party components. Components can install within the ASP data center and integrate as a bundled component of the enterprise application available to subscribers.

9

Portals and Intelligent Agents

Introduction

One of the major trends that continues to influence both organizational movement and networking technologies involves the application of knowledge management tools. Knowledge management covers the gathering, management, handling, controlling, organizing, dissemination, and application of information. In turn, knowledge management relies on Knowledge Management Processes that use technologies such as portals and intelligent agents for the production and integration of knowledge. All this leads to a planned and directed method of acquiring, producing, maintaining, improving, and transmitting the knowledge base that belongs to an organization.

Portals

Software applications called Enterprise Information Portals allow organizations to access stored information and to provide users a single gateway to personalized information. With this, the portal establishes a format for "one-stop" information shopping and includes shared services such as security, metadata storage, and personalization. In all scenarios, the portal provides a foundation for making decisions. The use of intelligent agents and software applications throughout a portal institutes methods for managing, analyzing, and distributing information within an enterprise and outside of the enterprise. To accomplish these tasks, enterprise information portals operate through a standardized Web interface.

The portals use intelligent agents to provide the capability to interact with customers and support the bidirectional exchange of information. As an example, customers have the capability to ask questions of a Web site or to share information as prompted by the agent. Most portal applications target content toward an organizational function and integrate different applications and stored data or content into a single system. Organizations can use the data and information acquired through the portal for further processing that might include marketing or production strategies.

eBusiness Information Portals

Electronic commerce, or eCommerce, describes the practice of selling over the Internet, and electronic business, or eBusiness, describes methods that use Web technologies to support all or most business practices. Practices such as customer relationship management, supply chain management, enterprise resource planning, knowledge processing, and knowledge management can better assist a business when coupled with the Web.

An eBusiness Information Portal, or eIP, uses Enterprise Information Portals, or EIP technology to support eBusiness processes that go beyond the enterprise. One type of eIP called the Extraprise Information Portals, or ExIP, supports an extended enterprise usually consisting of a community of trading partners. The community conducts business around a common enterprise that provides a value of mutual interest and hosts an Extranet. Because of the mutual interests of the host and the community, business occurs at a predictable and repetitive rate.

Another type of eIP called the Interprise Information Portal, or IIP, supports consortiums of independent companies that do not conduct business around a common network host. Given the independence of the members and the lack of a common host, business occurs at an unpredictable and irregular rate. Business transactions occurring through the IIP respond to individual demands on marketplaces that exhibit a mutual interest to the community.

Enterprise Knowledge Portals

Many organizations began working with Enterprise Information Portals because of possible competitive advantages and possible increases in returns on investment along with opportunities for innovation, greater productivity, increased effectiveness, and decreases in the cost of information. In addition, businesses continue to consider portals as a method for improving universal access to resources and providing a unified, dynamic, and easily maintained picture of data and information.

However, all these possibilities and opportunities depend on the quality and validity of information delivered through the portal. In contrast to an Enterprise Information Portal, an Enterprise Knowledge Portal, or EKP, employs knowledge production, knowledge integration, and knowledge management to ensure the quality and validity of information. In addition to providing information, the EKP also produces and manages information that checks the business information and data obtained through the portal. In brief, the EKP uses a variety of tools and methods to produce knowledge from accumulated information and to ease risks associated with decision making.

Rather than simply operate as a product of the World Wide Web, the Enterprise Knowledge Portal Web-enables the production, integration, and management of knowledge. To accomplish this, the EKP not only shares characteristic seen with EIP but also includes a personalized browser-based interface, a structured data management, and an unstructured content management. The EKP requires an integrated architecture that takes advantage of Knowledge Claim Objects, or KCOs.

Content Analysis and Content Management

Within the EKP, Content Analysis transforms unstructured content into data, information, or knowledge by describing the content in terms of attributes of media objects, attribute structures, and rules relating attributes. Content Management organizes, directs, and integrates content analysis and distribution efforts aimed at producing or distributing data, information, or knowledge.

With this in mind, a Content Management System acquires, processes, filters, analyzes, and distributes media objects that are contained in paper and electronic formats and have no structure. The system archives and restructures the media objects for easy retrieval and manipulation. Data, information, or knowledge resulting from the system processes becomes stored in either a centralized or distributed corporate repository.

Knowledge Claim Objects

Knowledge Claim Objects encompass knowledge claim data, or metadata that describe the validity held by the objects and methods that produce object behaviors. As a result, the EKP gains the capability to distinguish knowledge from information by providing information about the results of validity tests about any piece of information. In contrast, an Enterprise Information Portal does not track and store the validity of accumulated information. Therefore, the Enterprise Knowledge Portal produces metadata that has a much greater scope. When the Knowledge Claim Objects accumulate the validity information, the objects also record the full history of discussions and interactions and transform that information into knowledge.

The eXtended Markup Language provides content markup and metadata capabilities that create an environment for content analysis. Pieces of document content become treated as persistent XML-based knowledge claim objects. XML data in EKP sources provide a particularly convenient form of persisting object data and metadata that can pass in and out of distributed EKP components.

Meta-information

When the Knowledge Claim Object tests the validity of accumulated information and produces validity information, it also produces meta-information about the claim. In part, the meta-information also compares the specific knowledge claim against competing knowledge claims. With this feature, the meta-information provides a probability about the extent that an organization can rely on the target claim when compared to other claims.

In addition, the meta-information shows information about the strength of the target claim when compared to other competing claims. If the target claim has much strength, it begins to approach the level of organizational knowledge and provide more support for decisions. If the target claim has weaknesses, it begins to approach the level of false organization knowledge and provides little support for any decision making.

With the EKP supporting the full set of knowledge life-cycle activities, it also becomes a knowledge production tool. The tracking and storing of validation results within the Knowledge Claim Objects and the capability to separate strong knowledge claims from weak knowledge claims produces knowledge. Because new knowledge results from the application of EKP resources, information integration processes for the enterprise begin to orient toward the integration of knowledge, validity information, and business information.

The use cases implemented through the operation of the EKP must support the acquisition of information, individual and group learning, formulation and validation of knowledge claims, broadcast functions, search and retrieve functions, resource sharing, and teaching. In turn, each of these fundamental areas supports the planning, monitoring, and evaluating phases of decision cycles.

Artificial Knowledge Manager

Every Enterprise Knowledge Portal begins with a personalized desktop browser-based portal. Because the portal operates in conjunction with an integrative, logically centralized, but physically distributed Artificial Knowledge Manager, or AKM, it connects to all mission critical application sources and content stores within the enterprise. Distributed Artificial Knowledge Servers, or AKSs, and intelligent mobile agents make up the Artificial Knowledge Manager.

The distributed AKM balances processing loads across the enterprise and dynamically integrates the portal system when change occurs. The AKM intelligent agents manage all application servers, data and content stores, and clients in the enterprise. As a result, the Enterprise Knowledge Portal establishes a new work environment for enterprise knowledge workers. The new environment aligns with, supports, and partially automates individual and collaborative workflow by creating, distributing, and using data, information, and knowledge. The EKP system uses these resources to make and implement decisions and actions.

Intelligent Agents

An agent is an intelligent piece of autonomous software that interacts with its environment and with other agents while representing a user's interests. From a software development perspective, agents have evolved as a combination of distributed objects, active objects, business objects, and scriptable components. Because of these characteristics, agents work as modular, independently developed pieces of software that may combine to produce flexible and extensible application systems. As active components, software agents may have mobile properties, can react to specialized interfaces, remain driven by goals and plans rather than by procedural code, and have knowledge of the domain. All these attributes allow the agent to act on behalf of a user, to reduce complexity, and to free users from tasks.

An agent includes properties such as adaptability, autonomy, collaboration, intelligence, social ability, reactivity, proactivity, and mobility. Figure 9.1 illustrates the interaction between the properties. As shown in the diagram, a larger area coverage corresponds with greater agentlike characteristics. Table 9.1 provides a definition of the agent properties.

Intelligent agents have taken a place in a variety of information technology environments and work with different information systems. Tasks for intelligent agents include improving user interfaces, seeking personalized information from the World Wide Web, providing entertainment, and assisting with software development. Applications for intelligent agents cover management and administration, electronic commerce, information retrieval and management, messaging, and network management.

For example, software agents known as Web robots or softbots automatically travel across the Web, navigate through hyperlinks, and evaluate the content at each destination. From there, the agents categorize the data according to a taxonomic

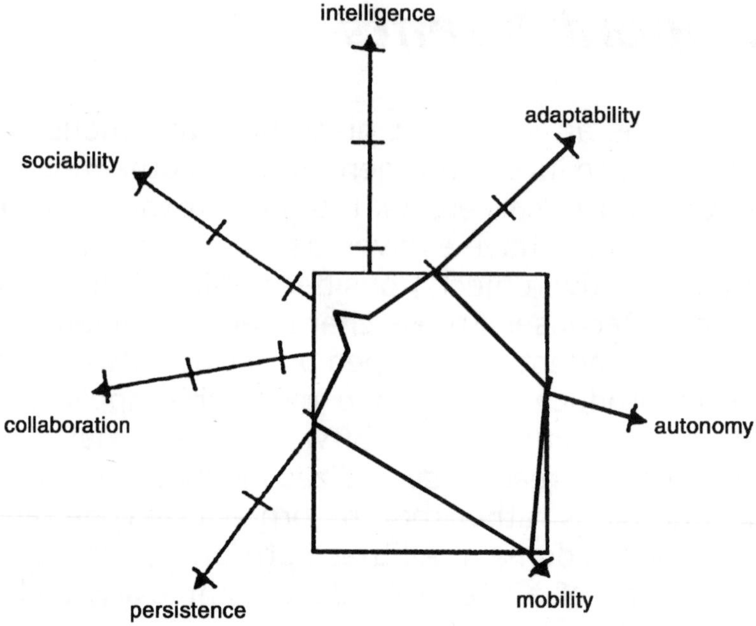

Figure 9.1. Interaction between agent properties

Adaptability	The degree that an agent' s behavior may be changed by downloading new programs, sending new rules, or otherwise changing the knowledge that the agent has by customizing behavior.
Autonomy	The degree of an agent's capability to execute one or more threads of control and to pursue some goal largely independent of messages sent from other agents. Agent autonomy differs from the methods of objects invoked by messages.
Collaboration	The degree of work cooperation and communication with other agents that forms multiagent systems working together on some task.
Intelligence	The degree of knowledge, understanding, and ability to reason from goals and knowledge displayed by an agent. Often an agent requires some artificial intelligence technology, such as rules, blackboards, neural nets, genetic algorithms, fuzzy logic, or knowledge bases to support intelligence.
Reactivity	The capability of an agent to perceive and respond to an environment.
Proactivity	The capability to respond to an environment, have goal-directed activity, and assume the initiative.
Social Ability	The capability to interact with other agents and with humans through an agent communication language.
Mobility	The ability of an agent to move from location to location. Mobility occurs either by moving the code and starting the agent afresh or by using complete serialization of code and state. Mobility allows the agent to begin execution at one location, complete work at the location, and then move to another location.

TABLE 9.1 — AGENT PROPERTIES

structure found within the client interface. Web robots can manage the integrity of the index database by automatically eliminating inactive links and adding links to current sites. Softbot tasks rely on user-defined parameters that initiate at the beginning of a search and continue so that the agent can work autonomously and provide updated information.

Given the broad range of intelligent agent tasks and applications, the technology base for the agents ranges from C++, Java, and scripts to neural networks. Despite this range, the development of an intelligent agent requires the consideration of several key issues. Intelligent agents must have a domain-specific vocabulary and a knowledge library that promote some type of reasoning capability. In addition to incorporating the reuse of communication and the capability to communicate with other agents or humans, intelligent agent development also requires a method of instruction so that agents can perform autonomous tasks.

We can separate agents into categories that include information integration, multiagent systems, assistant agents, social agents, mobile agents, and personal agents. As a group, agents can handle routine affairs, monitor activities, set up contracts, execute business processes, and find the best services for a customer. Using XML, agents can encode exchanged messages, documents, invoices, orders, service descriptions, and other information. Table 9.2 describes the different agent classes.

Agent-Based Software

Agent-based software development resembles conventional component development but leads to even more flexible modular systems that have less design time and implementation time and decreased coupling between the agent components. Agent component and agent systems use the same techniques that models, collaborations, interaction diagrams, pat-

Information Integration Agents	Have activities solely based on languages such as the agent communication language, the Knowledge Query and Manipulation Language, and the Knowledge Interchange Format standard.
Multiagent Systems	Rely on the Open Agent Architecture and work as part of Web-based commercial systems.
Assistant Agents	Include softbots and data-mining agents.
Personal Agents	Interact directly with the user and often present a sense of "personality" or "social skills." Personal agents can imitate an anthropomorphic character and have the goal of monitoring and adapting to the user's activities, learning the user's style and preferences, and automating or simplifying certain rote tasks. Research and development for personal agents involve animation techniques, natural language interaction, personalities, user profiles, and machine learning and intelligence.
Mobile Agents	Sent out to collect information or perform actions at one or more remote sites and then to return with results. They are typically implemented in Java or a portable scripting language such as TCL, Perl, or Python. Research and development for mobile agents include providing security, scalability, and fault tolerance; developing techniques to determine what capabilities agents can access at the remote site; establishing how to specify itineraries and interactions; and determining the appropriate levels of security and certification. Examples of mobile agent tasks include complex information access, telecommunications network management and service provisioning, and the other finding and aggregation of data pertinent to a particular query.
Social Agents	Communicate and interact as members of a multiagent system. Even though groups of agents represent users, organizations, and services, multiple agents engage in conversations as patterns of messages, to negotiate and exchange information. Research and development for social agents include work on intelligence, conversation management, agent-communication languages, ontologies, auction systems, and markets. Examples of social agent tasks include on-line auctions, planning, negotiation, eCommerce, logistics, supply chain, telecommunications, and system management. In most instances, several agents representing people and institutions work together to accomplish some business or information management goal.

TABLE 9.2 — AGENT CLASSES

terns, and aspect-oriented programming employ to help build more robust and flexible component and component systems. Consequently, agent-oriented programming offers a new method for considering the development of certain large complex distributed systems at a higher level of abstraction. This type of programming treats each agent as an autonomous entity that becomes driven by a set of beliefs, desires, and intentions.

XML and HTTP serve as easy-to-use, popular, heterogeneous, and platform neutral tools for agent development. Given the widespread usage of those tools, an array of robust and scalable implementations and other support exist. As with Java, XML and HTTP support Web servers, firewall access, and levels of security.

The following three major classes of technologies have begun to provide a basis for agents used to support electronic commerce:

- pervasive middleware such as HTTP and Web objects,
- business-to-business communication standards and models such as XML, and
- intelligent agents that take advantage of genetic algorithms, fuzzy logic, neural nets, rule-based systems, and agent technology.

Because XML has become the standard for data interchange on the Web, many programmers have chosen the language as the primary message format for dynamic agent communication. Dynamic agents send and receive information through XML-encoded messages, and XML tags mark up the information and break up the data into parts. In XML-based messages, agents encode information with meaningful structure and commonly agreed upon semantics.

In effect, an XML document works as an information container for reusable and customizable components available for use by any receiving agent. Making the Web accessible to agents with XML eliminates the need for customer interfaces for each consumer and supplier. As an example, a dynamic agent can carry an XML front-end to a database for a data exchange that involves XML encoded queries and answers.

COM, ActiveX, CORBA, JavaBeans, and Enterprise JavaBeans define slightly different component models that package software modules that have well-defined interfaces, distribution and composition properties, and a well-defined life cycle. The component provides some required interfaces that conform to the framework and applies many standard interfaces to access the support services. As a result of combining these services, the capability to conform to the framework makes component-based development a powerful and complex tool for dynamic agents.

A Typical Agent System

The term "agency" refers to a process or server that provides the local environment on a machine host for agents. The agency provides basic management, security, communication, persistence, naming, and — in the case of mobile agents — agent transport. A typical agent system provides the basic platform called an agent communication language, or ACL, and additional services that take the form of specialized agents. Figure 9.2 illustrates an overall agent system architecture that includes the components listed in table 9.3.

The agent management system and platform include capabilities for naming, communication, and conversation management. Naming identifies the agents for the purposes of sending the messages and changing their programs, rules, and knowledge. Agents may have a globally unique name or

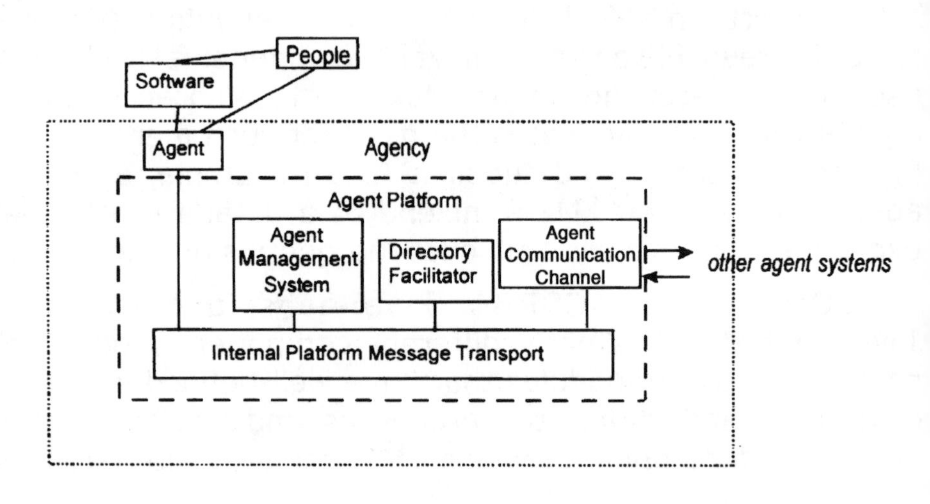

Figure 9.2. Representation of an overall agent system

Agent Management System (AMS)	Controls creation, deletion, suspension, resumption, authentication, and migration of agents. Provides local naming and locating of agents.
Agent Communication Channel (ACC)	Routes messages between local and remote agents, realizing messages using ACL.
Directory facilitator (DF)	Registers some agent capabilities so that an appropriate task-specific agent to handle the task can be found.
Agent Platform	Provides communication and naming infrastructure, using the Internal Platform Message Transport.

TABLE 9.3 — AGENT SYSTEM ARCHITECTURE COMPONENTS

may have one or more local names based on the agency or agent group. Even though a local or regional name may work well in most instances, mobile agents require a globally unique name that uses a distributed, hierarchical name service.

Communication provides a systematic and consistent way for agents to interact with each other, remain appropriately visible to the agent management system, and use an agent communication language, based on structured messages, known languages, and known vocabularies. In comparison, conversation management provides support for the monitoring and controlling of stylized message exchanges to ensure that the messages conform to a desired protocol or pattern. Infrastructure and agent services consist of the middleware, mechanisms, and services that support naming, communication, management, mobility, and persistence and legacy wrappers for nonagent software.

Agent Communication Languages

An agent communication language factors a communication problem into several independent parts that include the message type, addressing, context, and content or body of the message. Different ACLs factor messages in different ways and provide different standard parts. In all cases, the factoring has the goal of making the parts more reusable and the intent clearer and more actionable. Given the factoring of the problem into parts, an ACL enables more support from the agent and message management systems. As a result, the infrastructure can dynamically extend the agents to new problem areas while maintaining system checks on conformance to customer expectations.

Although the message type indicates the method for individually processing the messages and as part of a larger conversation, addressing indicates the sender and destination of the message. In addition, the message type indicates the key intent of the message. Context enables the understanding of the content of the message by establishing explicit language, vocabulary, and conversation patterns.

The content of the message may occur in languages that can accept XML encoding such as LISP, Prolog, TCL, Perl, JavaScript, or VB Script. Specifying the language defines the

Message Type	Used in conjunction with specified conversation protocols to allow the system to monitor and control the progress of conversations and confirm the compatibility with conventions and communication pragmatics. Also called performatives, communicative acts, verbs, or actions.
Addressing	Used to identify the sender and the intended receiver(s) of the message, using some form of local, regional, or global naming scheme. May also use originator for forwarded or brokered messages.
Message Sequencing	Used to connect a series of messages to a specific conversation and to each other, providing "in-reply-to" and "reply-with" identification.
Conversation Control	Used to connect a series of messages to a specific conversation. *Reply-by =date-time* sets a deadline for a timeout. *Protocol* identifies a specific type of conversation.
Vocabulary	Used to identify the domain of discourse to which the message content applies. Vocabulary defines objects and legal data types ("load"), action words ("adjust"), simple attributes, and more complex attributes. Also called ontology.
Language	Specifies the language in which the content is expressed.
Content	A statement, expression in the chosen language using terms from the chosen ontology.

TABLE 9.4 — ACL STRUCTURE PARTS

syntax of the content and semantics of keywords. However, the remainder of the content will refer to terms that relate primarily to the concepts in the communicated problem.

Most agent communication languages rely on a linguistic theory that studies how people communicate and engage in stylized conversations as a pattern. Typed messages indicate the intent of each communicative act. Moreover, the typed messages exchange according to stylized conversation patterns. Based on the type of the message sent, the ACL expects a certain response. The message types distinguish between commands, requests, information, and various kinds of responses such as acknowledgment or errors. With this system of differentiation, the ACL creates a basic structure that allows understanding of a conversation between agents to occur without examining the content of the message. Table 9.4 shows the parts of the ACL structure.

Scripting

Scripting languages are small, interpretive languages that have good access to the underlying component model such as VB Script or VBA for COM components, Java or JavaScript Java for JavaBeans, and TCL, Perl, or Python. A scriptable component must have the capability to expose additional interfaces. Developers and end users can create applications by using scripting to combine, control, or modify robust components created by domain experts.

Dynamic Agents

eCommerce establishes a distributed computing environment with dynamic relationships among a large number of autonomous service requesters, brokers, and providers. To match the automation required for efficient eCommerce operations, agents need to have dynamic behavior and the ca-

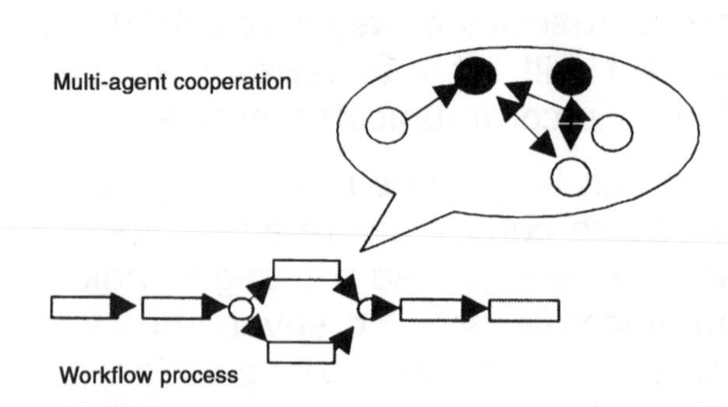

Figure 9.3. Multiagent cooperation in workflow

pability to maintain identity and consistent communication channels and to retain data, knowledge, and other system resources that may reach across applications. Dynamic agents provide the automation through a Java-coded, platform-neutral, extensible infrastructure. As pictured in figure 9.3, a dynamic agent has a fixed part and a changeable part.

Dynamic Agent Fixed Part

The fixed part of a dynamic agent provides a lightweight built-in management facility for distributed communication, object storage and resource management, action activation, and the graphical user interface, or GUI. Once initiated, a carried application becomes a reference to the underlying built-in management facilities and can use this reference to access the APIs of the facilities. A message-handler handles message queues and accommodates the sending, receiving, and interpreting of interagent messages through interaction styles that include one-way, request/reply, selective broadcast, and message forwarding. An action-handler handles the message-enabled execution of application programs.

Dynamic Agent Changeable Part

The data and programs carried by a dynamic agent form the changeable part. A dynamic agent has the capability to carry data, knowledge, and programs as objects and to execute the programs. Although all newly created agents have the same capabilities, application-specific behaviors occur and become modified through dynamically loading Java classes representing data, knowledge, and application programs. As a result, dynamic agents serve as general-purpose carriers of programs rather than of individual and application-specific programs.

One dynamic agent can carry multiple action programs. An open-server handler provides a variety of continuous services that can start and stop flexibly at a designated dynamic agent runtime. A resource handler maintains an object store for the dynamic agent that contains application-specific data, Java classes, and instances including language interpreters and addresses. Applications executed within a dynamic agent use the built-in dynamic agent management facilities to access and update application-specific data in the object store, and to perform interagent communication through messaging. Enabled by corresponding messages, a dynamic agent can load and store programs as Java classes or object instances and can initiate and execute the carried programs.

Dynamic Agent Functions

Within the programs carried by dynamic agents, built-in functions can access the dynamic agent's resources, activate and communicate with other actions run on the same dynamic agent. In addition, the functions can communicate with other dynamic agents or even stand-alone programs. Mobility allows the movement or cloning of a dynamic agent to a remote site and the sending of programs carried by one dynamic agent to another agent for execution at the receiving site.

Every dynamic agent has a unique symbolic name. A coordinator agent provides the naming service and maps the name of each agent to the current socket address for the agent. Operating as a dynamic agent, the coordinator also maintains the agent name registry and resource lists. Dynamic agents may form hierarchical groups that have a coordinator for each group and support a multilevel name service.

After the creation of a dynamic agent, the agent will attempt to register the symbolic name and address by sending a message to the coordinator. With the name registered, the newly created dynamic agent can communicate with other dynamic agents by name. When the new agent sends a message to another agent with an unknown address, the new agent can obtain the address from coordinator. If the new agent receives instructions to load a program but does not have an address for the loading, it can also obtain the address from the coordinator. Along with having the capability to obtain addresses from the coordinator, each dynamic agent also maintains an address book and records the addresses of known, active dynamic agents.

Moving to figure 9.4, note that a multiagent system requires both the naming service and other coordination. Given the needs of the multiagent system, the coordinator or other designated dynamic agents can provide coordination and brokering services. A resource broker maintains a hierarchically structured agent capability registry that includes leaf-level nodes referring to the dynamic agents that carry corresponding programming objects. Agents contact the resource broker when acquiring new programming objects or dropping new programming objects or when needing a program such as a domain-specific XML interpreter.

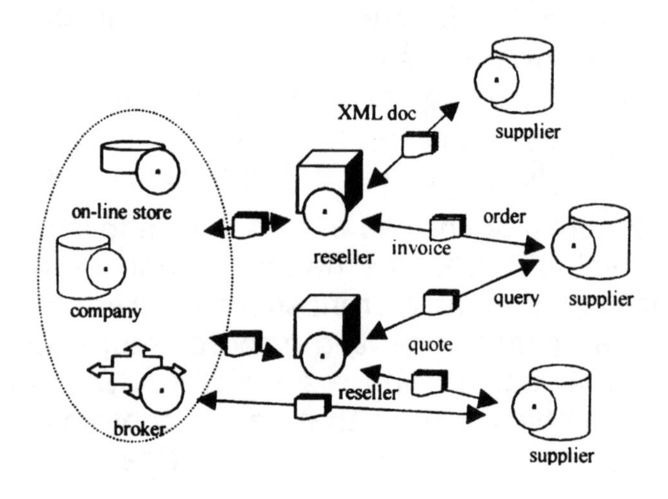

Figure 9.4. Multiagent cooperation with XML

A request broker isolates service requesters from dynamic agents that carry the services and allows an application to make requests for a service transparently. An event broker delivers events — treated as asynchronous agent messages — to event subscribers from event generators. Event notification may occur in two different ways. Point-to-point event notification provides a method where the event subscribers know the event generators and make the subscriptions accordingly. Multicast event notification provides a method where one or more event brokers handle events generated and subscribed anywhere in the given application domain. These event distribution mechanisms allow agents to subscribe to events without prior knowledge of their generators.

Dynamic Role Assignment and Switching

A role provides a dynamic interface between agents and cooperative processes. Agents playing a specific role follow the normative rules given by the cooperative process. Combined with the action-carrying capability, other capabilities of dynamic agents simplify mapping from a role to a specific agent and allow one agent to play different and sometimes simultaneous roles. A dynamic agent obtains the capability for playing a role by downloading the corresponding programs and passes a role to another agent by uploading the programs.

The agent can switch roles by executing different programs. To switch from one role to another, the agent must have memory capability across multiple activities and possess a consistent communication channel. As an example, an agent responsible for ordering a product may also have the responsibility for order cancellation. The action carrying and exchanging capability of dynamic agents support task reallocation. Each agent can contract out tasks if the task fits the functions of another agent or accept a task.

Multiagent Coordination

As illustrated in figure 9.5, multiagent systems involve static or dynamic groups agents working together to achieve some purpose and have a critical place in large-scale Ecommerce operations. The communication can occur between people and agents, between agent and agent, or in a hybrid format. Coordination covers the interactions between members of the group for the purpose of achieving a high-level task. As an example, a group of agents may have a series of stylized message exchanges to accomplish a task such as advertising and using a brokered service, bidding during an auction, responding to a call for proposals, or negotiating a price.

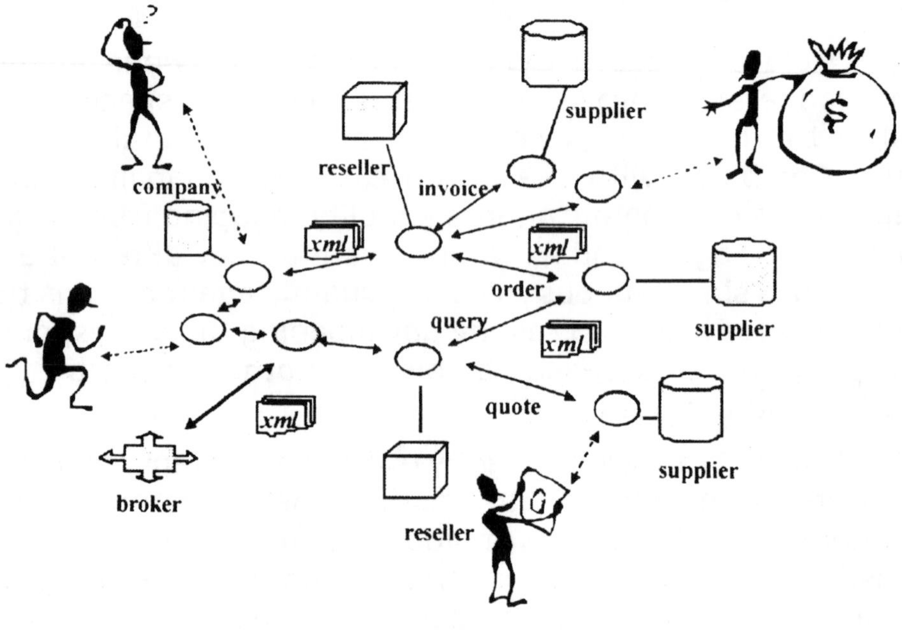

Figure 9.5. Illustration of multiple agents working in static and dynamic groups

10
Thin Client Networks

Introduction

Building on past lessons, thin clients create a robust computing environment that supports the expanding needs of medium and large businesses. Going back to mainframe technologies, users accessed programs and data through text commands typed into "dumb terminals" that consisted of a simple monitor and a keyboard. When mainframe technologies processed data over slow networks and used proprietary software, personal computers provided independent computing power at higher network speeds and through a graphical user interface.

Thin clients have become the next step in the computer and networking technology evolution. Operating as a networked desktop device that uses minimal local resources, a thin client establishes information access and delivery rather than information processing. Rather than relying on individual processing power, however, thin client computers take advantage of the power given through network servers.

Defining Thin Clients

The thin client consists of a network-connected computer that has minimal memory, disk storage, and processor power. In contrast to traditional networked computers — or fat clients — thin clients have a single point of administration at the server. Unlike fat clients that access the network to download program files and data files, thin clients exchange only small packets of information over the network. To a user, a thin client appears to operate much like the traditional desktop computers. However, thin clients offer additional features such as shadowing, location independence, and remote support.

Independent Computing Architecture

The Independent Computing Architecture, or ICA, display protocol transmits Windows GUI events across a network to connected devices. Because of its compatibility with Windows, ICA has become the most-used industry standard protocol for thin client systems. ICA provides universal access across any type of network to the full Windows NT operating system environment through the use of standard network connecting options such as a modem, Ethernet network, ISDN, Frame Relay network, or ATM network.

In addition, ICA allows Windows applications to run on a server while sending all graphical output to remote software or hardware for display. Working in the opposite direction, the ICA carries remote keystrokes and mouse events back to the application. Lightweight client software communicates with servers through the ICA.

Thin Client Operation

Most thin client networks rely on thin client-server software that allows access to existing Windows applications as well as internal custom applications. The networks allow users to run Windows applications remotely through server functions. Referring to figure 10.1, note that the server maintains all protocols and applications code. Because of this, organizations can preserve the investment in applications, receiving long-term cost savings from lower maintenance costs and cutting initial equipment purchase costs. Figure 10.2 shows a diagram of a thin client network that supports both personal computers and network computers.

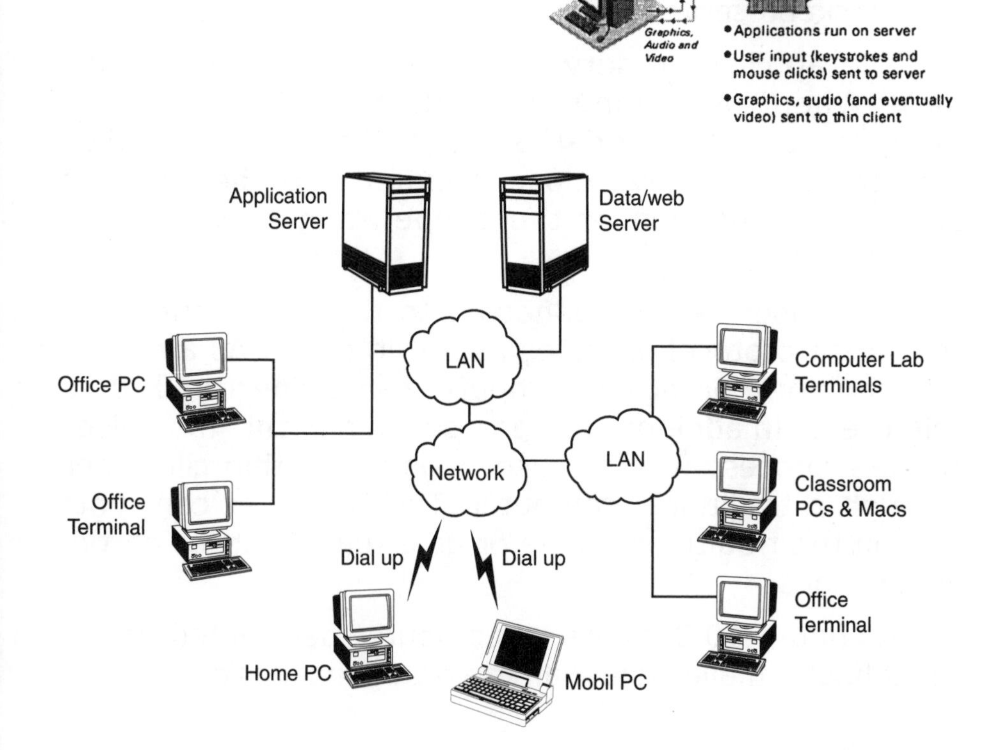

Figure 10.1. Thin client server contains all applications

Figure 10.2. Thin client network

The thin client solution includes the network infrastructure, the application server, a thin client device, and software. Because the network infrastructure serves as the most critical element in the thin client environment, a slow or inefficient network affects all users. Any organization moving to a thin client solution should begin with a cabling infrastructure that supports a minimum throughput of 10 Mbps. For transparent operation, the connections between the servers must provide high-speed data transportation.

In addition to the network infrastructure, the application server provides a robust and modular computer with a high fault tolerance. Again considering minimum configurations, the application server should have best-of-class processing power, offer 8-24 Mb of RAM per simultaneous user and maintain 64 Mb of RAM for the operating system. The application server should also provide miscellaneous server-side components used for managing applications and user environments such as load-balancing facilities for distributing client sessions across multiple backend servers.

The application server may work as part of a server farm. When considering the implementation of the server farm, servers used for data storage, Web hosting, and other services should have close proximity to the application server. The location of the cooperating servers reduces network traffic.

Any user device connected to the server through the thin client protocol operates as a thin client. As a result, almost any new or legacy computer can connect and run as a thin client. In addition, the phrase "thin client" may also describe a process where a user logs into a thin client server through a traditional computer. The computer can function both in the traditional mode or as a thin client for particular applications.

As figure 10.3 shows, a computer designated as a thin client has a smaller size than the traditional desktop computer

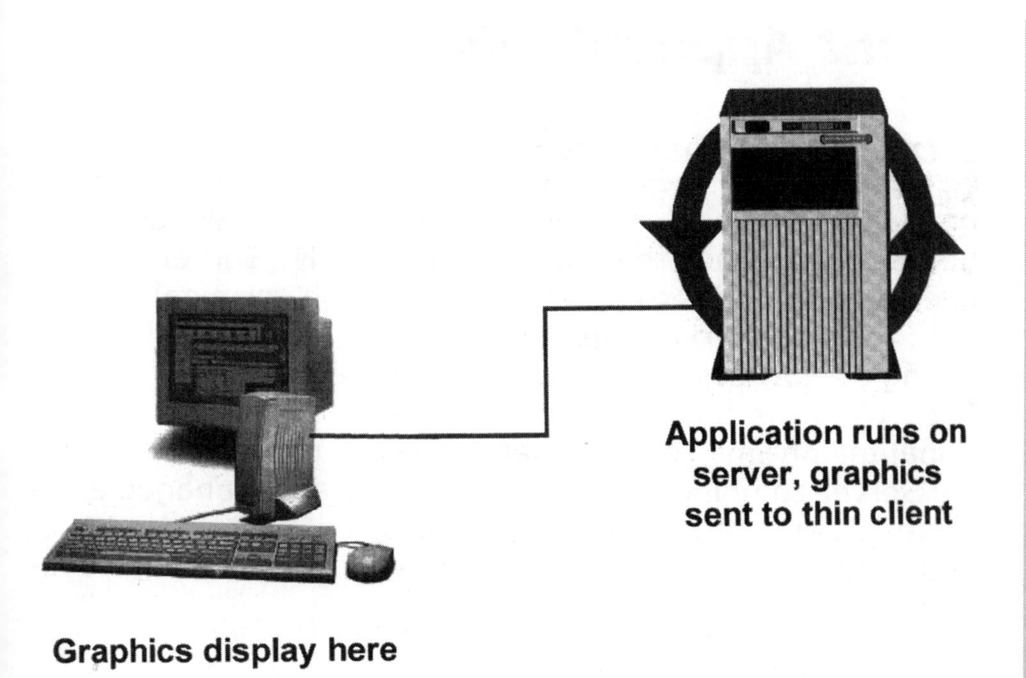

Figure 10.3. Thin client hardware

and may feature a sealed case without openings. A thin client computer requires only a processor capable of processing graphics, the capability to interface with an Ethernet network, a video card, and a minimum of 2 megabytes of RAM for running the thin client protocol. In comparison, thin clients do not rely on a hard disk drive, modem, floppy disk drive, or optical drive. Upgrades for the thin client occur through downloads from a manufacturer's Web site.

A thin client environment can take advantage of many software applications such as most Web-enabled or network-optimized productivity, graphic intensive, or specialized applications. Software licenses should cover an adequate number of simultaneous users. An application server requires either the Terminal Server Edition of Microsoft Windows NT server 4.0 or Terminal Services in Windows 2000 and Citrix MetaFrame.

Client Applications

Microsoft Terminal Server

The Terminal Server Edition of Microsoft Windows NT Server 4.0 extends the scalable Windows NT Server product line to diverse desktop hardware through terminal emulation. Terminal Server supports a full range of clients by combining the low cost of a terminal with the benefits of a managed Windows-based environment. In comparison to traditional mainframe terminals and desktop computers, Terminal Server combines low cost and a centrally managed environment with the familiarity, ease of use, and breadth of applications support offered by the Windows operating system platform.

Remote Desktop Protocol

The Remote Desktop Protocol operates as a key component of Terminal Server. Based on the International Telecommunications Union T-120 Protocol, the Remote Desktop Protocol provides a multiuser server core that offers the capability to host multiple, simultaneous client sessions. The protocol not only allows Terminal Server to host compatible multiuser client desktops directly on Windows and non-Windows-based hardware but also runs on Windows NT Server version 4.0 and above.

In addition, the protocol allows all standard Windows NT-based management infrastructure and technologies to assist with the management of the client desktops. The protocol facilitates communication between the client and Terminal Server over the network. The multichannel Remote Desktop Protocol supports three levels of encryption and high-bandwidth enterprise environments.

Terminal Server Client

Terminal Server Client displays the familiar 32-bit Windows User Interface on desktop hardware including Windows-based Terminal devices using embedded software running from the ROM. The client software also operates with personal computers running Windows 95, Windows 98, and Windows NT Workstation versions 3.51, 5.0, or 5.0. Client software also allows older personal computers that only have the capability to run Windows for Workgroups to operate as thin clients.

Terminal Server Administration Tools

Along with the familiar Windows NT Server administration tools, Terminal Server provides several new tools for managing client sessions. Those tools include the Terminal Server License Manager, Terminal Server Client Creator, Terminal Server Client Connection Configuration, and Terminal Server Administration tools. The Terminal Server software also adds two new objects called Session and User to the Performance Monitor to provide for the tuning of the server in a multiuser environment.

Citrix MetaFrame

Citrix offers MetaFrame as a server-based, multiuser software complement to the Microsoft Terminal Server Edition of Windows NT Server 4.0. MetaFrame incorporates ICA and extends Windows Terminal Server with additional client and server functionality. The Citrix product extends Microsoft Terminal Server with additional client- and server-side functionality in heterogeneous computing environments.

While Microsoft Terminal Server supports Windows-based devices and IP-based connections, MetaFrame delivers Windows-based application access to almost all types of client hardware, operating platforms, network connections, and LAN protocols. As a result, organizations can keep their existing infrastructure and provide access to the most advanced 32-bit Windows-based applications across the enterprise.

MetaFrame provides value-added functionality for all types of Windows clients including Windows 95, Windows CE, Windows NT Workstation, Windows for Workgroups, and Windows 3.x systems. In addition, MetaFrame also supports non-Windows clients including DOS, UNIX, Mac OS, Java, and OS/2 Warp and a broad range of client hardware including legacy PCs, Pentium PCs, Windows-based terminals, network computers, wireless devices, and information appliances. The Citrix product links to the network through standard telephone lines; T1, T3, 56Kb, and X.25 connections; ISDN, Frame Relay, and ATM connections; and wireless connections while supporting protocols such as TCP/IP, IPX, SPX, and NetBIOS.

MetaFrame enhances the user experience by providing complete access to all local system resources including full 16-bit stereo audio, local disk drives, COM ports, and local printers. Although applications run remotely from the server, the user works with an application that appears to run on the local drive. Establishing a local feel for the software increases the comfort level for users and lessens the need for training.

The additional functionality includes support for heterogeneous computing environments, enterprise-scale management, and seamless desktop integration. In addition, MetaFrame adds value to Terminal Server through its thin client-server add-on software. Moreover, MetaFrame not only transforms the deployment and management of business applications but also improves access to those applications. The software provides improved application performance and security.

Enterprise-Scale Management

MetaFrame includes robust, scaleable management tools that simplify the support of multiple applications and thousands of users enterprise-wide. The Citrix software also supports the easy and transparent addition of servers without any reconfiguration of the desktop computers. Once installed, MetaFrame allows the administration of applications across multiple servers from a single location. With MetaFrame software, organizations can cost-effectively manage and support the growth and complexity of large enterprises.

Systems Management Tools

Systems Management Tools provide greater manageability, scalability, and security and allow administrators to reduce the computing costs and resources needed to support users and systems. The Citrix Load Balancing Services facilitate the grouping of multiple MetaFrame and WinFrame servers into a unified server farm that meets the needs of a growing user base. As the size of the organization increases, administrators can add additional MetaFrame servers for more horizontal scalability. To protect network data, SecureICA Services offer end-to-end RSA RC5 encryption for the ICA data stream.

Application Management Tools

Application Management Tools simplify and accelerate the initial deployment and subsequent updates of applications across the enterprise. On one hand, Application Publishing utilities allow administrators to deploy applications easily across multiple MetaFrame and WinFrame servers from a single point. Application Launching and Embedding, or ALE, establishes the easy integration of Windows- and Web-based applications without the rewriting of code. The ReadyConnect Client accepts predefining with published applications, phone numbers, IP addresses, server names, and connection options. ReadyConnect Client is predefined prior to the first-time installation and facilitates the rapid, mass deployment of applications throughout the enterprise.

User Management Tools

To improve the productivity of end users and IT professionals, MetaFrame includes tools such as Session Shadowing and the Automatic Client Update utility. Session Shadowing allows administrators to assume control of a single user's or multiple users' sessions for support, diagnosis, and training. The Automatic Client Update utility allows the automatic updating of the Citrix ICA client software from the server and reduces the time and expense associated with the individual installation of client software.

Citrix WinFrame 1.8

WinFrame provides access to virtually any Windows application across any type of network connection to any type of client. Based on ICA technology, WinFrame offers a cost-effective and proven solution. WinFrame provides enterprise systems with centralized management, universal access, exceptional performance, and improved security for all business-critical applications and data. WinFrame also provides features for improving file, system, and application security. To improve the security of WinFrame servers on a corporate LAN, WAN, Intranet, or Internet, administrators can lock specific files, directories, and system areas as well as the entire systems.

Application Manager and Persistent Object Caching

The new WinFrame Application Manager provides support for single-point application publishing and allows secure, one-step application access for WinFrame users. Administrators simply click to assign applications to servers and to publish applications to users over LAN, WAN, and dial-up connections. To improve application performance and to reduce network traffic, Persistent Object Caching sends graphics to the client once and stores locally between application sessions. As a result, application splash screens and toolbars transmit once rather than download to the client each time.

Web-Based Application Launching and Embedding

A plug-in for Netscape Navigator and an ActiveX control for Microsoft Internet Explorer allows full-function Windows-based applications to launch from embedded HTML Web pages. From an appearance and performance perspective, the application seems to run on the local system, but it is actually executing on the server. Customers can provide on-line access to any existing application such as order entry, catalogs, groupware, or client databases over Intranets or the Internet without rewriting a single line of code.

By using the server-based application configuration utility, ICA file editor, and Application Presentation, Webmasters have complete control over the execution properties of the application and on-screen Web page presentation. Extremely secure operation in a Web environment succeeds because of

- a restricted application list that operates with a secure kernel to shut down any back doors that may exist in your applications,
- pre- and post-application execution scripts that can set up and clean up user environments, and
- a new government C2 security utility that triples the standard security levels provided by Microsoft in standard Windows NT.

Anonymous and Registered User Types

WinFrame supports two access types that define how users access a WinFrame server in a Web environment. Identification and password prompts challenge registered user types and ensure identification to the administrator. To deal with the public nature of the Internet, the system also offers anonymous user types not challenged with IDs and passwords. The anonymous user types share a Guest security level.

License pooling allows for the sharing of purchased WinFrame licenses across multiple servers within a server group, simplifies administration, and efficiently utilizes resources. In addition, license pooling does not require a key diskette for installation. Along with adding to ease-of-use, not requiring a key diskette enables network-based installations or the loading of a server across the network.

Protocol Compression Options

Various protocol compression options enable the turning on and turning off of ICA protocol compression based on available bandwidth. Active ICA protocol compression turns on for low-bandwidth dial-up and WAN connections and provides minimized bandwidth utilization that mimics the performance of a local-area network. For high-bandwidth LAN connections, deactivated compression reduces server load and improves processor utilization on the server while establishing the blazing performance of ICA.

Winterm Thin Clients

Winterm thin clients provide complete, enterprise-wide access to 16- and 32-bit Windows, Java, browser, and legacy applications with a user-friendly graphical interface. Along with plug-and-play capabilities, Winterm thin clients include integrated terminal emulation and an assortment of connectivity options for almost any enterprise network.

Winterm thin clients also have the capability to separate the interface of an application from the execution. During a session, only mouse clicks, keystrokes, and screen updates travel across the network. As a result, Winterm devices require only one-tenth the bandwidth seen with a conventional client-server network.

Thin client networks may include the Windows NT Server base operating system with extensions to NT's memory manager and scheduler, which provide NT with true multiuser capabilities. Server-based Thin Client Computing is taking a basic terminal and using a client-server model. It provides the capability to run all the software on the server and have just a terminal at the workstation.

Advancements in network technology made it possible to send larger and larger packets of information across networks at faster rates, and personal computers were networked to share resources and move data. As a result, the personal computers could function as both individual computers and clients of servers. However, the desktop control that added to the appeal of personal computers also added to the challenge to support the network and reduced communications capabilities between workstations.

Thin Clients and Application Service Providers

Application Service Providers offer an alternative to purchasing and supporting all four parts of a thin-client solution. As we learned in chapter 9, the ASP owns, tests, upgrades, and maintains software applications and server equipment. In addition, the ASP centralizes the cost and complexity of managing and delivering applications and serves those applications over secure Internet connections. When compared to the costs of implementing and maintaining the traditional thin client environment, working with an ASP requires less initial investment.

Working with the ASP

Project managers should review the applications currently used by the business and work with employees and clients to determine future needs that may require an ASP. After reviewing ASPs that offer appropriate service, they should select an ASP that has tested the applications and offers service bundles that contain necessary resources. Although most businesses purchase existing packages, the selected ASP may offer customization based on previous purchases or specialized needs.

Because your business depends on the quality of service obtained from the ASP, performance levels should be set in the contract. A service level agreement should be established that guarantees

- uptime,
- help desk response,
- a plan to remedy problems, and
- software upgrades.

The ASP should have a good security plan to protect data and applications.

As the project progresses, network and desktop device requirements should be analyzed and current and future needs should be determined. Based on the specifications given by the ASP, determine the type of upgrades needed for your current computing environment. Most ASPs require a T1 or other high-speed WAN connection and a LAN for clients. To ensure compatibility, ASPs also work with a consultant or specialist to review and test existing network, to verify existing documentation, to identify weaknesses, and to set performance benchmarks.

The majority of the work includes upgrading the network to meet bandwidth needs identified by the ASP and installing the thin clients. For network upgrades or installation, the consultant should develop a scope of work, timeline, equipment list, and a budget for the project. The scope of work might include

- upgrade of electrical, Ethernet wiring, or network closet facilities,
- network upgrades,
- thin client installation, and
- verification and testing.

Rather than estimate the cost of application upgrades and or additional server equipment, the budget should include a line item for the annual ASP subscription fee. An equipment list should encompass network components and wiring materials, whereas the budget should consider the installation and consulting costs of the project.

The installed thin clients should always be tested using a phased approach that begins with a few clients and adds more as a test of scalability and network throughput. Because the ASP provides the applications and determines the configuration settings, the installation of thin clients usually entails nothing more than plugging in and booting up. The ASP should resolve network-related issues such as slow response times and application view problems. Although the ASP will remotely handle server maintenance and upgrades, the network connections remain the responsibility of the business.

Thin Client Benefits

Thin clients offer businesses a reliable and secure alternative to the costs associated with implementing traditional network technologies. Cost savings occur through the optimization of network resources, providing access to applications from older computers that may rely on either the Windows, Macintosh, or UNIX operating systems, keeping data secure and uncorrupted and providing technology staff with more control. Other benefits include shadowing and the capability to maintain machine independence.

Shadowing

Shadowing allows certain users access to another user's desktop in real time. For example, a corporate trainer can assume control of a student's desktop and methods for utilizing application tools. If a student becomes confused or lost, a trainer can access the desktop remotely and demonstrate the problem solution. Although a few applications on the market offer similar features, the thin client software performs this function at the server rather than the desktop. With an ASP configuration, only the system administrator can use the shadow function.

Machine Independence

Because the desktop configuration establishes according to user log-in rather than the device, any device can access the server. Users see the same icons and buttons and have access to the same applications whether using the application from the office, home, or remote location. Information and work stored on a centralized server becomes instantly available from any location and available to appropriate groups according to log-in privileges.

Designing a Thin Client Environment

Businesses have successfully deployed thin clients as a solution to the escalating cost of ownership. Thin client network designs begin with the centralization of computing power in application servers and a high capacity network. Many times, businesses will rely on a systems integrator, consultant, or Value-Added Reseller that has experience in designing and implementing thin-client networks. A Value-Added Reseller, or VAR, should

- have at least 2 years of experience with thin client installations,
- operate as an authorized Citrix partner,
- employ staff members who have both Microsoft Certified System Engineer Citrix Certified Administrator credentials,
- provide documentation as part of the finished project, and
- define user groups according to privileges and applications set by the local system administrator.

Depending on the scope of work defined, an integrator assists at several levels of the project such as support with planning, design, and evaluation of existing equipment. Integrators also serve as the primary contact with thin client equipment and software vendors while offering experience important to identifying the best combination of products for the system. System administration for thin servers requires specialized training for the configuration of user groups, addition of new users, addition of applications, and development of a backup and restoration process. The integrator should offer clear documentation and appropriate training.

Installing Thin Clients

In cooperation with the project manager, an integrator should review the current network, servers, and computers for the purpose of determining the best method for integrating a new system into the existing one. The review should also cover the replacement or usage of existing equipment. A safe, secure location for servers should have appropriate heating, air conditioning, and ventilation.

Analyze Requirements and Determine Needs

A single thin client environment may include 30 to 3,000 thin clients. Equipment requirements depend on network users, access requirements, and applications. The first step defines the primary groups of users and access requirements. Users may have a unique login with permission set according to task level or other group attributes. Because some applications operate better than others on thin clients, determine which specialized software may require fat clients.

Consideration of the applications used on the thin client network should include the potential number of required concurrent applications and the potential number of concurrent users. Server utilities can track and measure application usage. Remote management software can be used to run diagnostics and set benchmarks for service. The software also identifies signaling thresholds to prompt a review of the service before users notice problems.

Always test applications to determine compatibility and bandwidth needs. As an example, many applications written for Windows 3.1 or later and DOS remain compatible with the thin client protocol. Graphic-intensive applications may require more bandwidth than other applications. The testing of servers should occur in a controlled environment and on a non-critical part of the network. Test all the applications and user configurations while simulating actual business-day situations.

11

Wireless LAN and PAN Technologies

Introduction

With the release of an Ethernet-equivalent standard, wireless networking technologies have begun to gain widespread deployment. In turn, the installation and implementation of wireless networks has led to the accelerated development of low-cost, interoperable products. Along with networking devices such as adapters, Access Points, and gateways, new devices such as Internet phones and hand-held computers have complemented the usefulness of the network. Wireless technologies now cover devices such as notebook computers, personal digital assistants, Internet access appliances, and Voice-over-IP telephones.

Until recently, wireless networking had taken a vertical focus on industry uses. This early adoption of wireless technologies included retail and warehousing with usage of handheld devices covering tasks such as data collection and inventory management. With the introduction of the IEEE 802.11, Bluetooth, and HomeRF standards, the tasks associated with wireless networks have also introduced businesses and schools to a more flexible approach to adding networks, personal computers, and peripherals to the workplace.

Because the mobility associated with wireless network technologies improves productivity and service, the wireless systems can provide LAN users with access to real-time information anywhere in their organization. As a result, wireless networking can allow a transition to tasks such as the delivery of patient information between doctors and nurses, the training of new employees, the exchange of information with central databases, classroom activities, and consulting activities.

Wireless LANs can augment rather than replace wired LAN networks by providing the final few meters of connectivity between a backbone network and the in-building or mobile user. In addition, the use of wireless technologies can reduce the cost of network ownership by minimizing the overhead that accompanies changes within a network. Along with eliminating the need for installing cables, wireless networks simplify many installation and configuration issues. The portability and scalability of wireless networks allows network managers to preconfigure and troubleshoot entire networks before installation at remote locations proceeds.

Defining Wireless Local-Area Network Technologies

As depicted in figure 11.1, a Wireless Local-Area Network, or WLAN, operates as a flexible data communications system that can either replace or extend a wired LAN. Even though a WLAN uses many of the same technologies seen

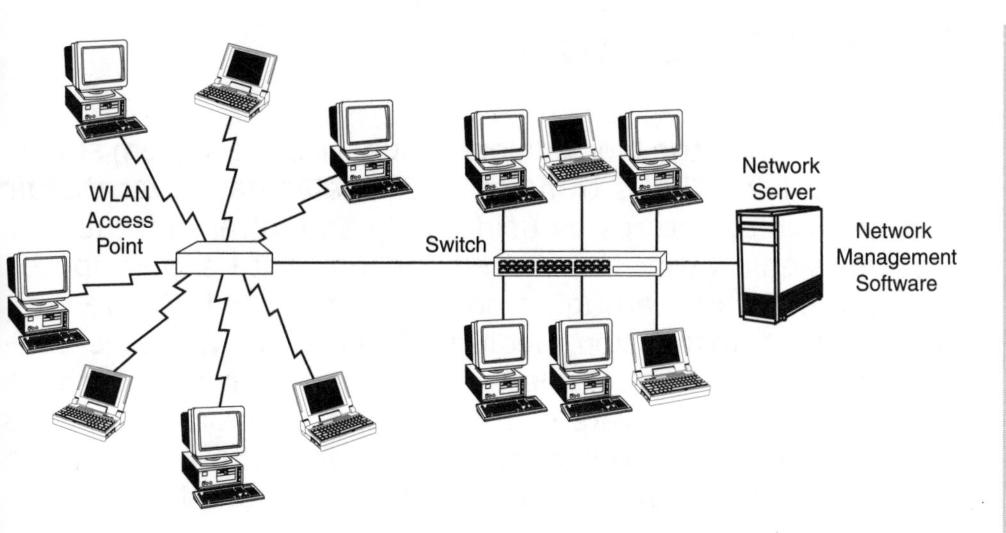

Figure 11.1. Illustration of WLANs and wired network interaction

with broadband wireless networks, packet radio systems used by law enforcement or utility units, and cellular or packet radio, the WLANs systems differ considerably. In comparison to WLANs, the broadband wireless, packet, and cellular systems offer much lower data rates. In addition, users usually pay for bandwidth and services on a time or usage basis. WLANs technologies do not just rely on a less costly infrastructure, WLANs also offer much faster data transmission rates and do not require usage fees.

Because WLANs provide all the features and advantages seen with traditional local-area network technologies, WLANs offer more functionality and flexibility. The features associated with WLANs technologies have begun to redefine the meaning and capabilities of network infrastructure. The functionality and flexibility offered by wireless networking technologies has combined with the acceptance of the IEEE 802.11 High Rate WLANs standard by vendors who produce network operating systems, application software, and network hardware. To ensure cross-vendor interoperability, the Wireless Ethernet Compatibility Alliance, or WECA, certifies products and places the Wireless Fidelity, or Wi-Fi, symbol on products compatible with the standard.

WLAN Operation

As shown in the block diagram of a wireless system shown in figure 11.2, WLANs transmit and receive data over the air and through structures by using radio and infrared frequency transmissions rather than wired cabling. Data superimposes onto a radio wave through modulation; the carrier wave then acts as the transmission medium. A carrier wave delivers information in the form of energy to a remote receiver. Because the frequency or bit rate of the modulating information adds to the carrier, the modulation of data onto the carrier wave causes the radio signal to occupy more than one frequency.

Several factors affect the distance for radio frequency, or RF, signal communication. Along with the receiver design, the amount of transmitted power and the propagation path can change the design of a wireless network. Walls, metal, and even people can affect how energy propagates and — as a result — the range and coverage that a system design can achieve. Other factors include the number of users, the existence of multipath interference, the quality of the WLANs devices, system latency, and bottlenecks within the wired network.

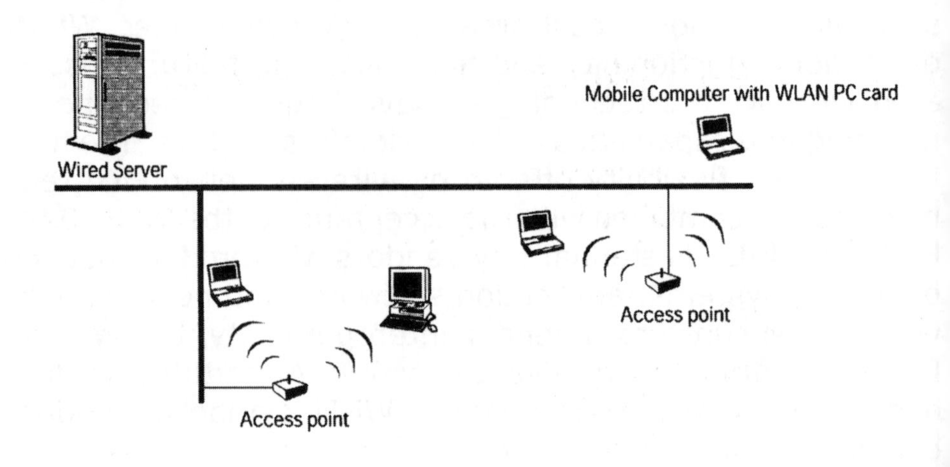

Figure 11.2. A Wireless LAN configuration

As Figure 11.3 illustrates, a radio signal can take multiple paths from a transmitter to a receiver. With multipath, signal reflections may either strengthen or weaken signals and affect data throughput. Due to longer path lengths, these reflected or refracted signals take longer to arrive at the receiver. When the signals arrive at the signal through the multiple paths and at different times, interference with the main signal occurs. Multipath characteristics depend on the number of reflective surfaces in the environment, the distance from the transmitter to the receiver, the product design, and the radio technology.

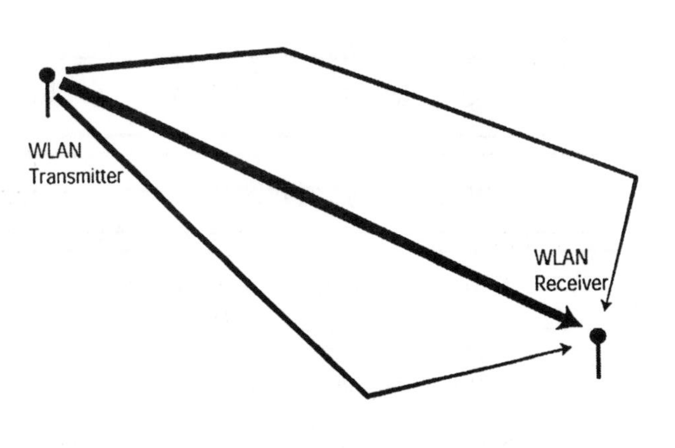

Figure 11.3. Radio signals traveling over multiple paths

Most WLANs operate within either the 2.4- or 5-GHz frequency bands. Multiple wireless users operate at the same time without interference through the use of different frequencies at the carrier. To extract data, a radio receiver selects one radio frequency while rejecting all other radio signals on different frequencies.

Wireless Application Protocol

Referring to the framework shown in figure 11.4, the Wireless Application Protocol, or WAP, serves as the wireless equivalent to TCP/IP. Subsets of WAP include the Wireless MetaLanguage, or WML, and Wireless Transport Layer Security, or WTLS. WML serves the same role as HTML but applies to a restricted GUI environment that may include low-resolution or a monochrome display. WTLS certificates are 40 percent smaller than the standard X.509 certificates used in wired networks.

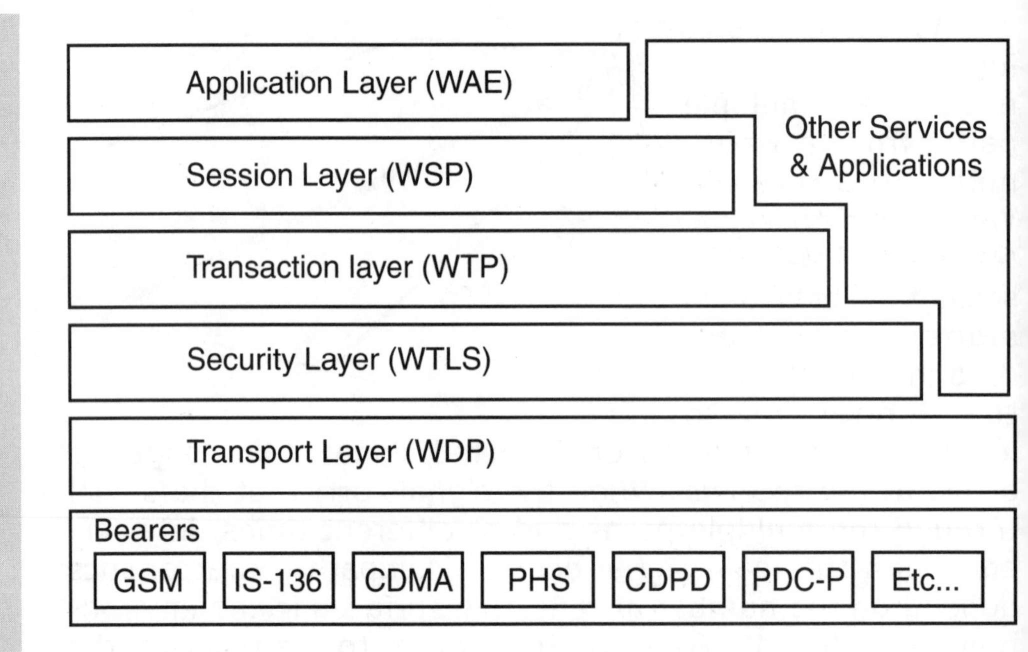

Figure 11.4. Wireless access protocol

WAP addresses the constraints of the wireless world by reducing the bandwidth and processing complexity of wired protocols. To accomplish this, WAP uses a proxy server called the WAP gateway that sits at the edge of the wireless network, bridges the gap between wireless and wired networks, and coverts data protected using WTLS into data protected by the secure socket layer. In each instance, WAP achieves significant bandwidth savings.

Along with those advantages, the Wireless Application Protocol brings Internet content and telephony services to digital cellular telephones and other wireless terminals. WAP reuses upper software applications that can interact with applications on the personal computer. The use of application gateways to mediate between WAP servers and PC applications allows the implementation of remote control and transfer of data from a computer to a handset.

Wireless LAN Hardware

Three main pieces — adapters, Access Points, and bridges — form the basis of the Wireless Local-Area Network. With this, WLANs address the transmission of RF signals, the baseband transport of the data, reception of the data at a modem/baseband converter, medium access control, and memory. The necessary receiver sensitivity for the wireless transmission of data at the higher rate requires lower bit-error rates and low circuit losses. Routing of the WLAN signals must ensure that the synthesizer or local oscillator does not read frequencies into the signal.

Referring to figure 11.5, note that the network adapters used to connect computers to the WLANs have the same appearance and function as the PCMCIA, Cardbus, PCI, and USB network adapters that operate with wired networks. In a wired Ethernet or Token Ring LAN, the adapters establish access to the network by providing an interface between the network operating system and the cabling. In a Wireless LAN, the adapters establish a transparent connection to the network by providing an interface between the network operating system and an antenna.

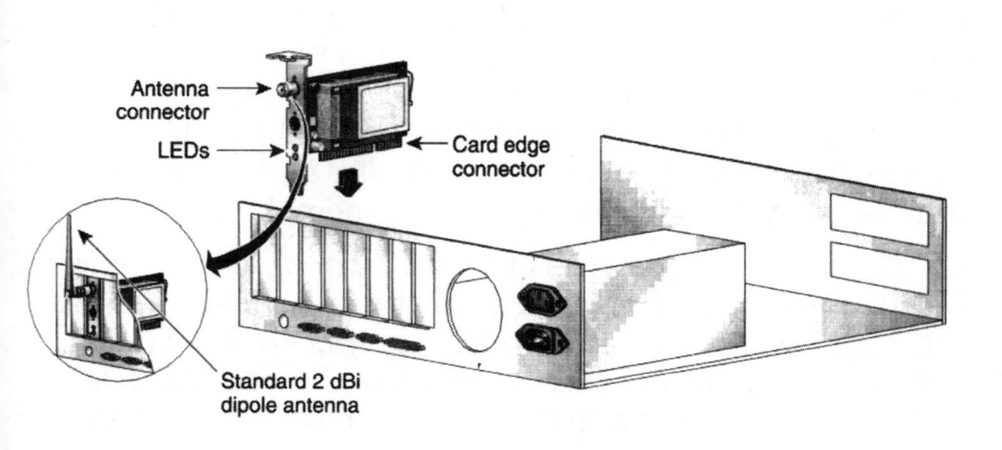

Figure 11.5. Installation of a WLAN adapter

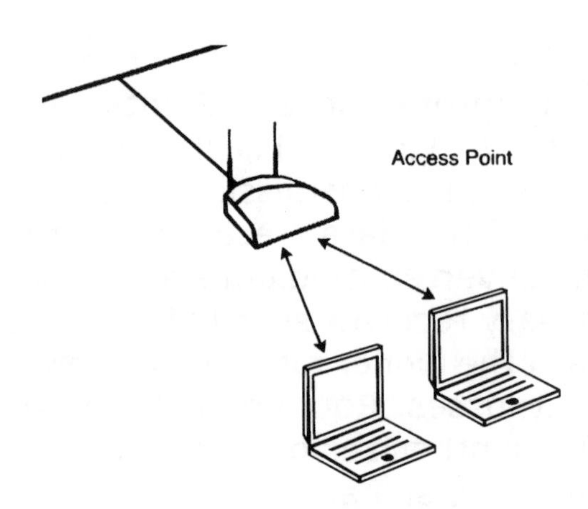

Figure 11.6. The Access Point receives, buffers, and transmits data between the wired network and the WLAN

In a typical WLAN configuration, a transceiver — or Access Point — connects to the wired network from a fixed location using standard Ethernet cable. As such, an access point operates as the wireless equivalent of a LAN hub. Moving to figure 11.6, note that the access point receives, buffers, and transmits data between the WLAN and the wired network while supporting a group of wireless user devices. Access points connect with the wired backbone through a standard Ethernet cable and communicate with wireless devices through an antenna.

Either the access point or the access point antenna mount high on a wall or attach to a ceiling. Because the given power output for a signal limits the distance of a wireless communication, WLANs use microcells to extend the connectivity range. Each access point has a range that extends from less than 20 meters to 500 meters. As figure 11.7 indicates, multiple access points operate by handing off data from one to another as the user moves from location to location. At any point in time, a mobile computer equipped with a WLAN adapter associates with a single Access Point and microcell. Individual microcells overlap to allow continuous communication within a wired network and handle low-power signals by "handing off" users as they roam through a given geographic area.

Depending on the type of WLAN technology, the network configuration, and the network applications, a single access point can support between 15 and 250 users. WLANs offer easy scalability because more access points can be added as needed. The addition of access points decreases network congestion and enlarges the coverage area. Generally, large facilities that re-

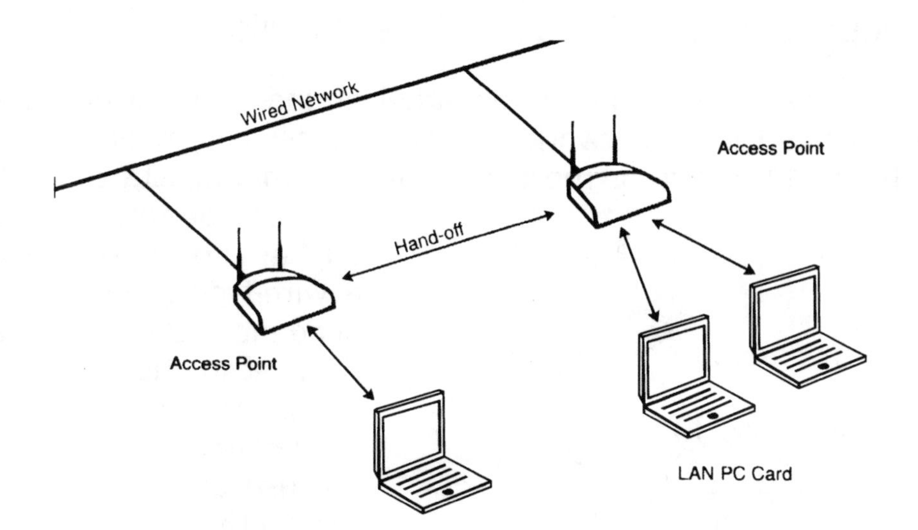

Figure 11.7. Handoff from Access Point to Access Point

quire multiple access points overlap the cells for constant connectivity to the network. A wireless access point can track the movement of clients across its domain and permit or deny specific traffic or clients from communicating through it.

Point-to-point local-area wireless solutions — such as LAN-LAN bridging and personal-area networks, or PANs — may overlap with some WLAN applications. However, those solutions address different user needs. A Wireless LAN-LAN bridge provides an alternative to the cable that connects LANs in two separate buildings. A Wireless PAN typically covers the few feet surrounding a user's workspace and provides the capability to synchronize computers, transfer files, and gain access to local peripherals.

Outdoor LAN bridges connect WLANs located in different buildings. Compared to the cost of constructing or replacing a current cabled infrastructure, the cost associated with a WLANs bridge may seem attractive. The bridge can also provide a method for a receiver to detect a narrowband signal by either controlling the rate that the signal moves from one frequency to another or by placing a unique code within the signal.

Independent and Infrastructure WLANs

A WLAN may configure as either an independent network or an infrastructure network. With the independent model shown in figure 11.8, two or more computers communicate through wireless adapter cards but have no connection to a wired network. Independent WLANs provide the best solution for situations that do not require an infrastructure through the use of peer-to-peer connectivity. Any time two or more wireless adapters remain within range of each other, the users can establish an on-demand, independent network that does not require administration or preconfiguration. Because Access Points can act as a repeater, the use of access points within an independent network can effectively double the distance between the nodes.

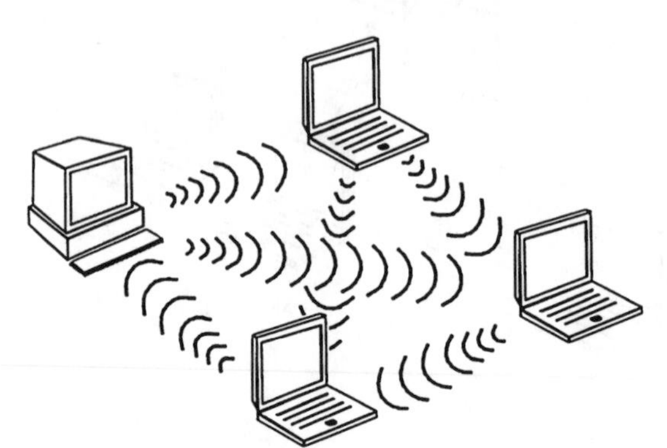

Figure 11.8. Independent Wireless LAN

An infrastructure WLAN establishes a client-server environment by offering fully distributed connectivity through

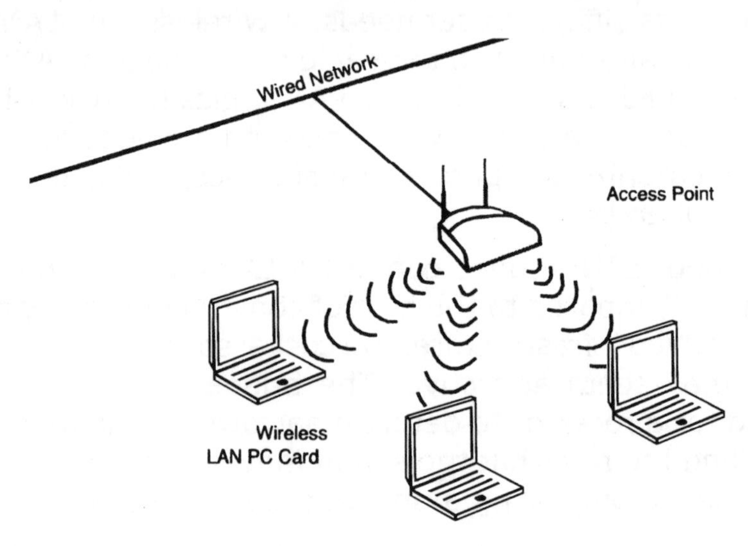

Figure 11.9. Client-server wireless configuration

multiple computers using a wireless adapter card to communicate with a central hub. Multiple Access Points link the WLAN to the wired network, allow users to share network resources efficiently, and can provide wireless coverage for an entire building or campus. The Access Points not only provide communication with the wired network but also mediate wireless network traffic for the immediate location. Figure 11.9 shows a diagram of an infrastructure WLAN.

Defining Personal Area Network Technologies

A Personal-Area Network supports the communication and synchronization of personal mobile devices such as personal digital assistants, pagers, palmtops, laptops, and other smart appliances. The communication between the devices and between the devices and standard fixed computers on the enterprise network includes the delivery of voice, data, Web applications, video clips, and graphics. Synchronization includes tasks such as synchronizing the personal contacts list between a mobile phone, notebook computer, and a hand-held device.

As illustrated in figure 11.10, personal-area network applications range from the type of ad hoc connectivity associated with file transfer to the use of cordless computer interfaces such as wireless mice, keyboards, game pads, and joysticks. In addition, PAN applications include the use of cordless peripheral devices, access to wired LAN applications, and localized Wireless LAN access. Individuals connected to a PAN have the capability to check Web sites, download electronic mail, complete file transfers, synchronize devices with corporate servers, and access mobile telephones.

Because of the need for commonplace personal-area network access, manufacturers have moved toward embedding the radio link and firmware into a single integrated device.

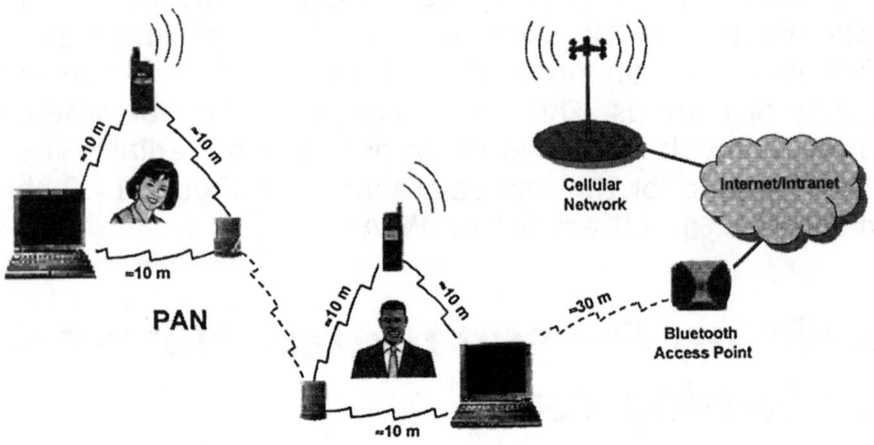

Figure 11.10. Illustration of personal-area network interaction

Aside from the complexity involved with the need to provide common devices, additional complexity surfaces because of specific applications for personal-area networks. Even though home networking uses plug-and-play capabilities and the automatic registry of devices to the network, using PAN devices for business applications requires additional security. The addition of multimedia capabilities for the devices also adds to the need for sufficient bandwidth and protocols to handle the transfer of files.

IEEE 802.15 Personal-Area Network Standards

Established during 1999, the IEEE 802.15 committee has begun to establish personal-area networking standards. The standards support point-to-point and multipoint connections between devices and transmission power levels. In addition, the 802.15 committee has addressed the frequency ranges for personal-area network devices as well as the compatibility between the devices and equipment attached to either a wireless or wired network.

Spread Spectrum Transmission

As the name indicates, a narrowband radio system maintains a radio signal frequency that has sufficient bandwidth to only pass information. A narrowband radio system transmits and receives information on a specific radio frequency and avoids undesirable crosstalk between communications channels by carefully coordinating different users on different channel frequencies. The radio receiver filters out all radio signals except the ones on its designated frequency.

A spread spectrum system spreads the transmitted signal over a frequency much wider than the minimum bandwidth required for sending the signal. The spread spectrum design consumes more bandwidth but produces additional reliability, integrity, and security through a louder and easier-to-detect signal. If the receiver does not tune to the correct frequency, the spread spectrum signal resembles background noise.

The Federal Communications Commission does not require licenses for operation at the 2.4-GHz band. Consequently, the popularity of the band has begun to encompass other devices that do not transmit energy within the same frequency spectrum but can also interfere with the operation of a WLAN or PAN. Although the use of the spread spectrum does not solve all interference issues, the use of spread spectrum methods improves the stability and reliability of connections.

Frequency-Hopping Spread Spectrum

Referring to the illustration in figure 11.11, note that the Frequency-Hopping Spread Spectrum, or FHSS, method changes the frequency according to a predetermined rate acknowledged by both the transmitter and receiver. To accomplish this change, FHSS splits the available spectrum into separate bands. As the Access Point and client hop from band to band, portions of the data transfer during each hop. Systems

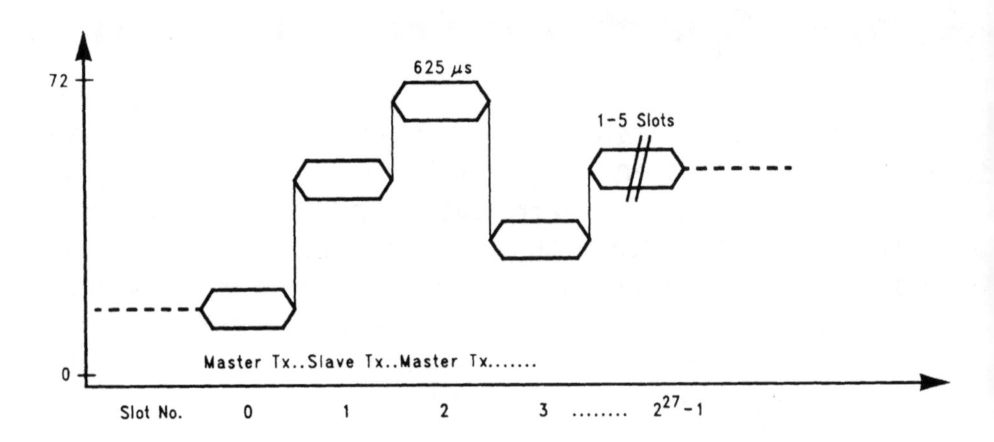

Figure 11.11. Diagram of frequency hopping

based on FHSS avoid interference with other transmission signals in the same band by hopping over many different frequency channels. When interference interrupts the data transfer because of transmitters colliding on a given frequency, the devices resume the transfer with the next hop to a new frequency.

However, the FHSS method of hopping between frequencies and reestablishing a connection adds to the overhead associated with transmitting the signal. Although the connection does not break, bandwidth decreases with each blocked frequency. Compared to the Direct Sequence Spread System method, the Frequency-Hopping Spread Spectrum offers greater power efficiency.

When properly synchronized, the transmitter and receiver monitor the changing frequency and maintain a single logical channel. To an unintended receiver, the FHSS transmission appears similar to short-duration impulse noise. FHSS systems have become popular for low-power, lower bit rate, low-range applications such as 2.4-GHz cordless phones and do not interoperate with Direct Sequence Spread Spectrum products.

Direct Sequence Spread Spectrum

As shown in figure 11.12, the Direct Sequence Spread Spectrum, or DSSS, generates a pseudorandom noise code that has a redundant bit pattern for each bit set for transmission and spreads the signal across a wider bandwidth. A DSSS system transmits data, multiplying the message with the pseudorandom nose code before transmission occurs. The broadening of the signal over a wider bandwidth adds to the capability of the system to avoid multipath, noise, and interference.

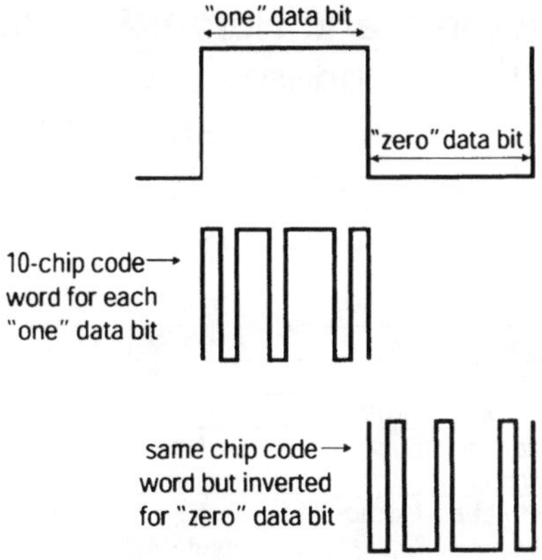

Figure 11.12. Direct sequence spread spectrum

Due to higher processing efficiency, DSSS devices achieve a slightly higher throughput than FHSS devices. Again comparing the two methods, DSSS offers greater range and support for transmission in the 5-GHz band. The spreading of the DSSS signal creates a lower power density across the spectrum. In a multiple-user installation, each transmission spreads with a different pseudorandom noise code and transmits within the same frequency band. At the receiver, the DSSS signal multiples by the appropriate pseudorandom code.

With each bit of data encoded into 11-chip sequences, DSSS protects the data against noise interference. Referred to as a chipping code, the bit pattern provides some reliability through length. A longer bit pattern establishes a greater probability for recovery of the original data. Although a longer bit pattern requires additional bandwidth, statistical techniques embedded in the radio can recover the original data even if one or more bits in the chip have become damaged and without retransmission. An unintended narrowband receiver sees the DSSS signal as low-power wideband noise and rejects the signal.

Wireless LAN and PAN Standards

Three different wireless standards have provoked interest in the promise of wireless technologies. The IEEE 802.11 standards target professional and Wireless LAN applications, and Bluetooth and HomeRF target consumer applications. The latter two standards use many of the components seen with the 802.11 standard, and both have reduced power requirements, transceiver capabilities, and functions. As shown in table 11.1, all three standards rely on spread spectrum technologies.

Specification	IEEE 802.11	Bluetooth	HomeRF
Physical Layer	FHSS, DSSS, IR	FHSS	FHSS
FHSS Hop Frequency	1 or Mbps	1600 hops per second	50 hops per second
Data Transmission Rates	11 Mbps	1 Mbps	1 or 2 Mbps
Modulation Method	FHSS - Two- or four-level gaussian frequency shift keying DSSS - Differential binary phase shift keying and differential quadrature phase shift keying	Shaped, binary frequency modulation	Two-or four-level frequency shift keying
Support for Data Security	40-bit RC4	0-, 40-, 64-bit encryption	Blowfish
Support for Multiple Devices	Maximum of 26 collocated networks	Multiple devices	Maximum of 127 devices in the network
Transmission Range	400 feet indoors 1,000 feet outdoors	10 meters unamplified 100 meters amplified	50 meters

TABLE 11.1 — COMPARISON OF IEEE 802.11, BLUETOOTH, AND HOMERF SPECIFICATIONS

The IEEE 802.11 Standard

During 1997, the IEEE adopted the 802.11 business Wireless LAN standard that supports 1- and 2-Mbps data rates in the 2.4-GHz band. The standard uses Frequency Hopping Spread Spectrum, Direct Sequence Spread Spectrum, and infrared physical layers. To comply with out-of-band regulations, the standard uses a 2-MHz guard band at the lower edge of the spectrum and a 3.5-MHz guard band at the upper edge.

The Federal Communications Commission specifies that the 2.4-GHz band can have a maximum of 83 FHSS frequencies and a maximum of 3 DSSS channels. Because two users on a channel must share the same bandwidth, the possible total throughput for the system decreases. With the FHSS method allowing more channels in the same frequency band, FHSS can support more total bandwidth for coverage.

Given the use of the point-coordination protocol scheme, the 802.11 standard supports synchronous data and voice or asynchronous data. A carrier sense protocol called the Distributed Coordination Function, or DCF, handles asynchronous communication. Supported by all 802.11-based wireless products, DCF provides nondeterministic transmission of information packets. Figure 11.13 represents the 802.11 packet structure.

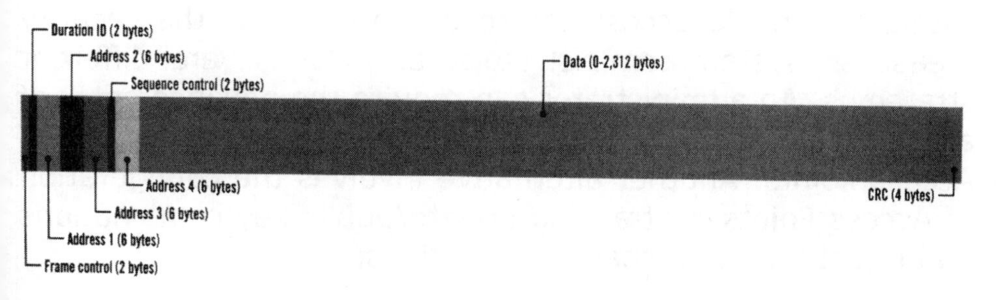

Figure 11.13. 802.11 packet structure

With nondeterministic transmission, no method exists for tracking the length of time needed for a data packet to move through the network. Similar to Time Division Multiple Access, or TDMA, the Point Coordination Function, or PCF, guarantees that the client owns a certain amount of bandwidth and supports synchronous transmission. Compared to DCF, PCF meets the needs of broadband wireless transmissions and provides better support for time-bound transmissions such as video and system control functions.

802.11 Security

Given the openness of radio-based communications, security concerns exist. The 802.11 standard incorporates a shared-key encryption mechanism known as Wired Equivalent Privacy, or WEP. During operation, WEP uses a series of challenges and verification schemes to secure wireless data. When a client attempts to connect to an Access Point, the Access Point sends a challenge value to the station. After the client receives the challenge value, the client uses the shared key to encrypt the challenge and send it to the Access Point for verification.

Although WEP uses 40-bit encryption, some 802.11b equipment manufacturers either offer optional 128-bit encryption, plan to make the higher level security available as a firmware upgrade, or manufacturer wireless NICs that have unique MAC addresses and a unique public/private key pair. The 802.11 standard provides an option of wired equivalent privacy by allowing the user to embed the RSA RC4 security algorithm within the media access controller. To establish the security measures, system administrators can follow several different strategies. An administrator can require the advance entry of all allowable hardware address/public key combinations into Access Points. Another alternative involves the configuration of Access Points for tracking private/public key combinations and rejecting unmatched combinations.

The IEEE 802.11a Extension

The 802.11a extension uses the 5-GHz band rather than the 2.4-GHz band and relies on an Orthogonal Frequency Division Multiplexing, or OFDM, scheme. In contrast to the IEEE 802.11b extension, IEEE 802.11a relies on only a Direct Sequence Spread Spectrum physical layer. Moving to figure 11.14, note that the extension includes a maximum of eight channels at the 20-MHz spacing within the lower 200 MHz of the 5-GHz

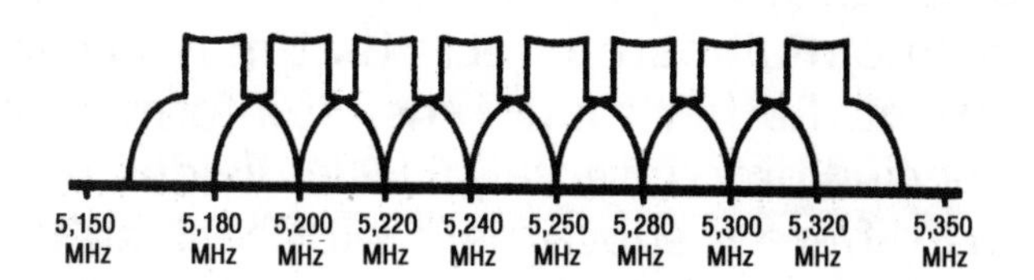

Figure 11.14. Clear channels in the 5-GHz spectrum

spectrum and four channels in the upper 100 MHz of the spectrum. As a result, the 5-GHz band provides four times the bandwidth seen with the 2.4-GHz spectrum.

The 802.11a extension provides optional turbo modes at a maximum of 54 Mbps and requires mandatory support of the 6-, 12-, and 24-Mbps data rates. The optional turbo modes offer support for new multimedia network technologies such as the multiple streaming of MP3 audio and Voice-over-IP telephony, Internet access, digital television, MPEG-2 DVD streams, and MPEG-4 Video-on-Demand. Implementation of the turbo modes becomes more difficult because the modulation scheme changes from binary phase shift keying and quadrature phase shift keying to 16-quadrature amplitude modulation for 24 Mbps and 64 quadrature amplitude modulation at speeds greater than 24 Mbps.

Orthogonal Frequency Division Multiplexing

As illustrated in figure 11.15, Orthogonal Frequency Division Multiplexing transmits multiple user symbols in parallel using different subcarriers and provides a more robust transmission scheme than seen with DSSS or FSSS. Although the subcarriers have

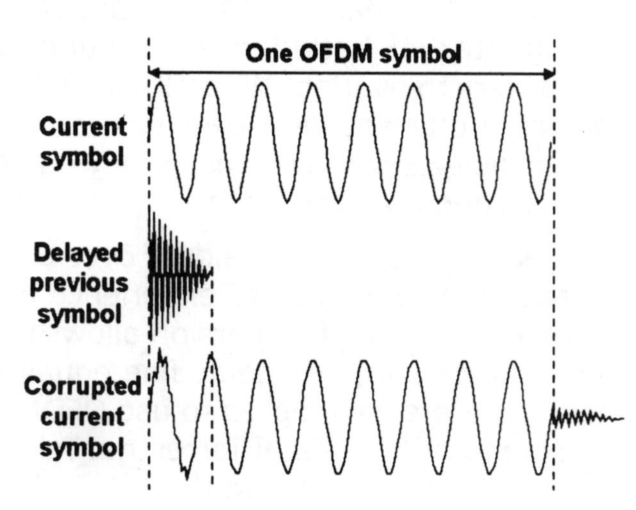

Figure 11.15. Orthogonal frequency division multiplexing

sync-shaped spectra, the signal waveforms remain orthogonal and overcome multipath effects that may occur in a business environment. Rather than transmit information over one carrier at a specific time interval, OFDM divides the transmission among 52 different subcarriers with each subcarrier having a transmission time lengthened by N.

Figure 11.16 shows the separation of the 20-MHz channels into 52 subcarriers. Despite the fact that the data rate for each individual subcarrier is reduced by a factor of *N*, the paralleling of 52 different transmissions results in a consistent transmission rate. In addition, each subcarrier has *N* times more multipath tolerance.

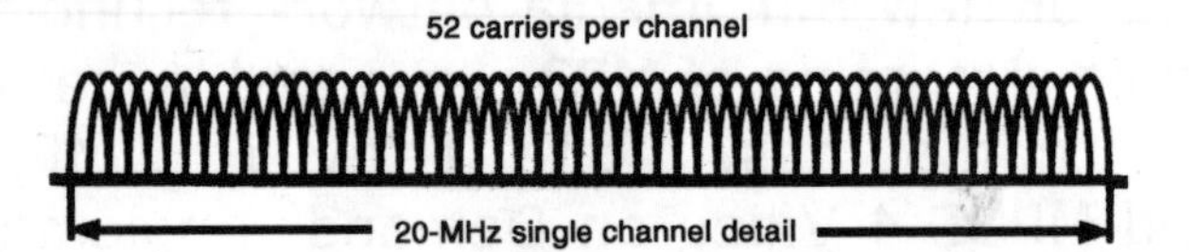

Figure 11.16. Subcarriers in the OFDM 20-MHz channels

Compared to other modulation methods, OFDM uses symbols that have a relatively long time duration and narrow bandwidth. Because each subcarrier experiences frequency-flat fading because the small bandwidth remains small enough in bandwidth to experience frequency-flat fading, the subcarriers remain orthogonal when received over a frequency-selective but time-invariant channel. As a result, OFDM does not rely on a guard interval between each subcarrier and allows the subcarriers to overlap. With all this, OFDM stands as the most spectrally efficient data transmission methods. The modulation technique can transmit a large amount of data within a given frequency bandwidth.

Receiving the OFDM signal over a time-invariant channel causes each subcarrier to experience a different attenuation. However, the lack of dispersion allows OFDM systems to avoid the need for a tapped delay line equalizer. As a result, many manufacturers have opted to use OFDM signals for Digital Audio Broadcasting, Digital Terrestrial Television Broadcasts, and Wireless Local-Area Networks.

The IEEE 802.11b Extension

Recently ratified by the IEEE as a high rate standard for WLANs, the 802.11b specification uses Ethernet-like protocols to provide up to 11 Mbps throughput with fallback rates of 5.5, 2, and 1 Mbps. At 1 to 2 Mbps, IEEE 802.11b supports a distance of 400 feet; at 11 Mbps, the range decreases to 150 feet and 20 dBm. The second extension uses a complementary code keying waveform.

Like all IEEE 802 standards, the 802.11b standard operates at the Physical Layer and Data Link Layer of the OSI model. An Ethernet-like Link Layer Protocol combines with acknowledgments of packets to provide reliable data delivery and efficient use of network bandwidth. Any LAN application, network operating system, or protocol will run on 802.11 compliant WLANs. Given the speed and range of 802.11b, users can run all but the most bandwidth-intensive network operations.

The IEEE 802.11b standard also includes features such as dynamic rate shifting, load balancing, and the simultaneous support of voice and data communications. When RF conditions begin to deteriorate, dynamic rate shifting allows the network to fall back to the lower data transmission rates. If an Access Point becomes congested with traffic or provides a low-quality signal or load balancing, the adapter cards change the associated Access Point to improve performance. The 802.11b standard also supports simultaneous voice and data communication.

In addition to the optional use of the shared key RC4 algorithm, several WLAN vendors provide access control features that authenticate clients and access points before a user may gain entry to the network. The 802.11b High Rate standard also supports two power-utilization modes called the Continuous Aware Mode and Power Save Polling Mode. With the Continuous Aware Mode, the radio remains activated and draws power. With the Power Save Polling Mode, the radio enters a sleep state that extends the battery life of portable devices. When a device uses the Power Save Polling Mode, the associated access point holds data in buffers and signals clients who have waiting traffic.The DSSS transmitting and receiving process despreads the desired received signal back to the original transmitted data while increasing the spread of all other signals and noise not correlated with the pseudorandom noise code. However, IEEE 802.11b standard DSSS systems can only tolerate interference up to half the strength of the desired signal. In practical terms, this tolerance translates into less than typical interference presented by one user to another user on the same channel. Although DSSS allows the independent sending and receiving of an individual signal in the presence of other desired signals within the same frequency band, the lower tolerance for interference in the 802.11b DSSS standard decreases the capability to share the spectrum.

Bluetooth

Bluetooth technology provides a wireless personal-area networking solution that can coexist with the WLANs standards while establishing device-to-device connectivity such as synchronization, short data transfers, or voice pass-throughs. As with the IEEE 802.11 standard, Bluetooth devices operate within the 2.4-GHz radio spectrum. The devices provide a maximum data transfer rate of 1 Mbps without line-of-sight requirements and maximum ranges of 10 meters client-to-client in open air, 5 meters in building, 100 meters client-to-Access Point in open air, and 30 meters within a building. Referring to figure 11.17, note that Bluetooth devices use the Gaussian Frequency Shift Keying, or GFSK, modulation scheme where a positive frequency deviation represents a binary one and a negative frequency deviation represents a binary zero.

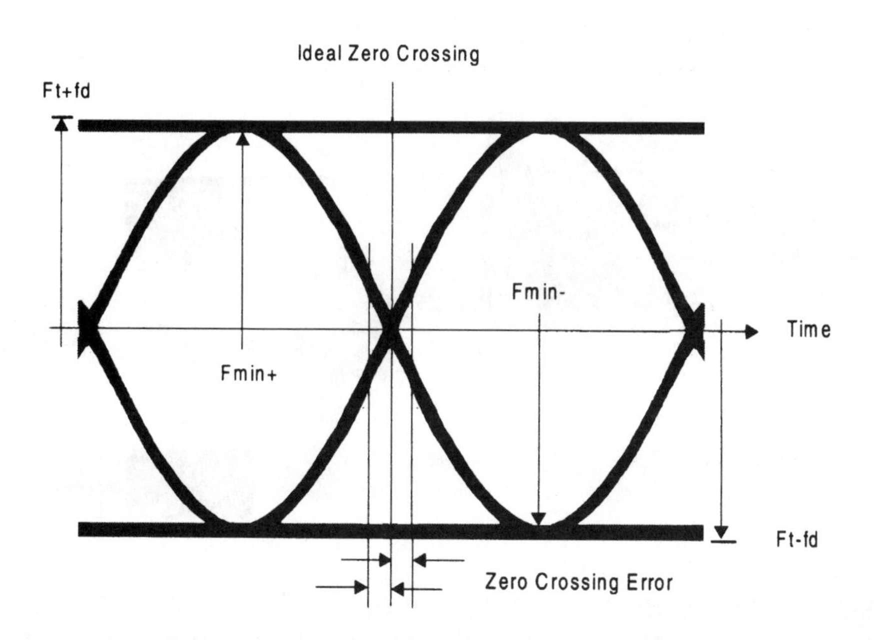

Figure 11.17. Waveform of the Bluetooth modulation scheme

Bluetooth devices support the transmission of voice and data through three voice channels and seven data connections per personal-area network. Originally developed as a solution for targeted point-to-point short-range links for voice applications, Bluetooth functions have expanded to include connecting the personal network work area and distinct personal-area networks. Bluetooth devices can form ad hoc networks with a combination of devices such as cellular phones linked to a Private Branch eXchange through Access Points and gateways. Because more than 1,300 companies support the Bluetooth standard, the technology provides interoperability and the automatic discovery of other Bluetooth devices and Bluetooth applications on other devices. Bluetooth eliminates the need for dependency on common operating systems and devices through the use of standard Bluetooth profiles that connect application to application.

With this, Bluetooth has the capability to establish interoperability with all mobile computing and communications devices and to ensure connectivity to the Internet regardless

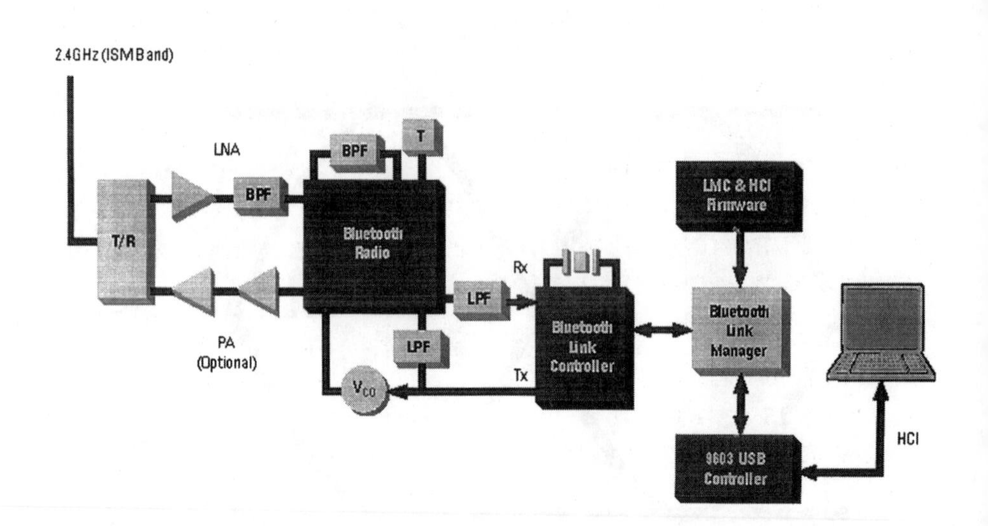

Figure 11.18. Block diagram of a Bluetooth adapter attached to a network computer

of size, power budget, platform type, and line-of-sight restrictions. The Bluetooth specification defines a lower power level that covers the shorter personal area within a room and a higher power level that can cover a medium range, such as within a home. Software controls and identity coding built into each microchip ensure that only those units preset by their owners can communicate. Security for Bluetooth devices occurs through integrated 0-, 40-, and 64-bit encryption and authentication.

Figure 11.18 shows a block diagram of a Bluetooth transceiver that attaches to a notebook computer through a USB port. Bluetooth devices have three general types of applications:

- PAN applications where two or more client products with Bluetooth directly communicate with one another.
- LAN access applications where a Bluetooth client communicates to the broader network through a Bluetooth LAN Access Point.
- WAN access applications where a Bluetooth client communicates through a Wireless WAN device to gain connectivity.

In addition, personal devices that include embedded Bluetooth radios can use a Bluetooth access point as a gateway to a corporate network or the Internet for low-bandwidth access. As hand-held devices and Web-enabled phones become more popular, users will gain the capability to take advantage of application gateways for simple and fast access to networked information through Bluetooth access points while roaming from location to another.

The Bluetooth wireless technology supports both point-to-point and point-to-multipoint connections. A maximum of seven slave devices can form a piconet while communicating with a master radio in one device. The linking of several piconets together in ad hoc scatternets allows communication to occur among continually flexible configurations. Although all devices in the same piconet have priority synchronization, other devices can enter at any time after receiving the proper configuration.

Bluetooth Protocols

Figure 11.19 shows the complete Bluetooth protocol stack. Rather than having all applications make use of all the protocols shown in the figure, applications run over one or more vertical slices from this protocol stack. Additional vertical slices support services of the main application such as the Telephony Control Specification or Service Discovery Protocol. The complete protocol stack is composed of both Bluetooth-specific protocols such as the Link Management Protocol and Logical Link Control and Adaptation Protocol, or L2CAP, along with non-Bluetooth-specific protocols such as the Object Exchange Protocol, or OBEX, and User Datagram Protocol.

The Bluetooth protocol stack divides into four layers according to purpose. Referring again to figure 11.19, note that the Bluetooth core protocols include the LMP, L2CAP, SDP, Cable Replacement Protocol (RFCOMM), telephony control protocols, and the adopted protocols. In addition, the Bluetooth Specification defines a Host Controller Interface, or HCI, that pro-

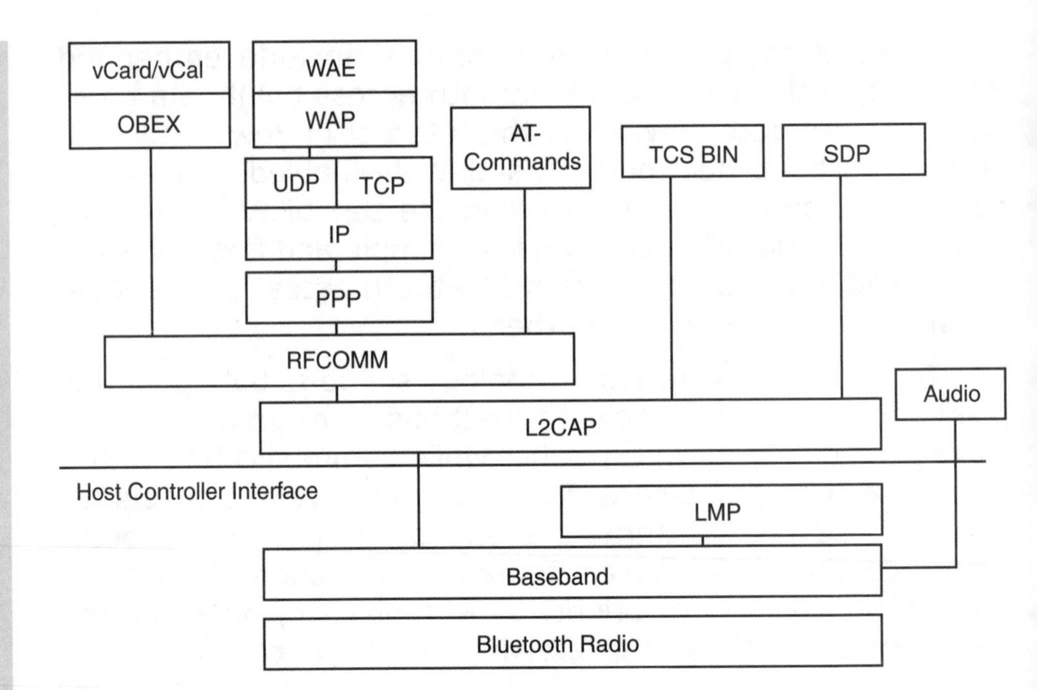

Figure 11.19. Bluetooth Protocol stack

vides a command interface to the baseband controller and link manager and access to hardware status and control registers. The Cable Replacement layer, the Telephony Control layer, and the Adopted Protocol layer combine to form application-oriented protocols that enable applications to run over the Bluetooth core protocols.

The Baseband and Link Control layer enables the physical RF link between Bluetooth units to form a piconet. Because of the use of FHSS in the Bluetooth system, the Baseband and Link Control layer uses inquiry and paging procedures to synchronize the transmission-hopping frequency and clock of different Bluetooth devices. The layer also provides the Synchronous Connection-Oriented, or SCO, and Asynchronous Connectionless, or ACL, physical links that can transmit as multiplexed data the same RF link. ACL packets only operate with data, but the SCO packet can contain audio only or a combination of audio and data. The Link Manager Protocol manages the link setup between Bluetooth devices through the control of power modes, duty cycles, and the control and negotiation of baseband packet sizes. In addition, the LMP includes security tools such as authentication and encryption by generating, exchanging, and checking link and encryption keys. The Logical Link Control and Adaptation Protocol adapts upper layer protocols and provides services to the upper layer when the payload data does not send through LMP messages. L2CAP provides connection-oriented and connectionless data services to the upper layer protocols with protocol-multiplexing capability, segmentation and reassembly operation, and group abstractions. Moreover, L2CAP permits higher level protocols and applications to transmit and receive L2CAP data packets up to 64 kilobytes in length.

The Service Discovery Protocol, or SDP, forms a crucial part of the Bluetooth framework by providing a basis for all the usage models. Using SDP device information, Bluetooth can query characteristics of the services and establish a connection between two or more Bluetooth devices. The Cable Replacement Protocol emulates RS-232 control and data signals over Bluetooth baseband and provides both transport capabilities for upper level services. Along with defining the call control signaling needed for the establishment of speech and data calls between Bluetooth devices, the Telephony Control Protocol defines mobility management procedures for handling groups of Bluetooth TCS devices.

The main principle behind the Bluetooth protocols maximizes the reuse of existing protocols for different purposes at the higher layers. Protocol reuse adapts existing applications to interoperate with the Bluetooth technology. In addition, the open system approach taken by the Bluetooth designers allows vendors to implement their own proprietary or commonly used application protocols freely on top of the Bluetooth-specific protocols.

As already mentioned, Bluetooth also takes advantage of several adopted protocols. In the Bluetooth technology, the Internet Engineering Task Force Point-to-Point Protocol runs over RFCOMM to accomplish point-to-point connections. In addition, a socket programming interface model allows the implementation of the TCP/UDP/IP protocol standards. The implementation of these standards in Bluetooth devices such as a cellular handset or data Access Point allows communication with any other device connected to the Internet:

The OBEX Protocol operates as a session protocol and provides a method for exchanging objects. Comparable to HTTP, OBEX uses a client-server model and remains independent of the transport mechanism and transport API. Along with presenting a format for conversations between devices, OBEX also provides a model for representing objects and operations. The OBEX Protocol defines a folder listing option used to browse the contents of folders on remote device.

Bluetooth Security

The Bluetooth specification includes security features at the link level and supports unidirectional or mutual authentication and encryption. Bluetooth bases the security features on a secret link key shared by a pair of devices. A pairing procedure generates the key when the two devices communicate for the first time. Referring to figures 11.20 and 11.21, note that the encryption engine initialization occurs after the system load and before the actual transmission or reception of data. The encryption sequence includes 208 initialization bits

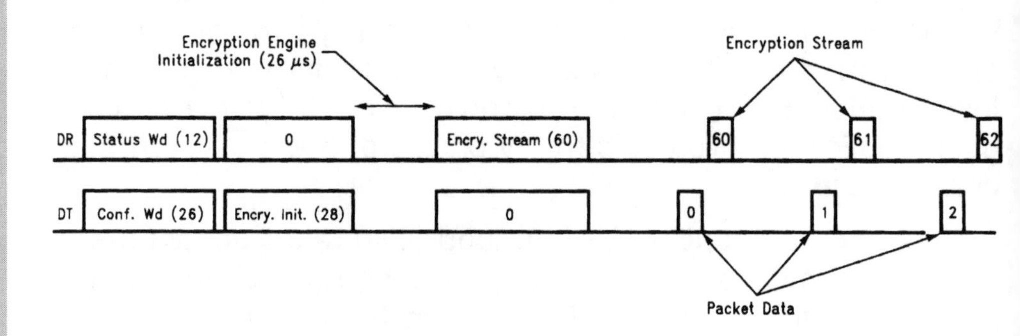

Figure 11.20. Transmitting Bluetooth data with encryption enabled

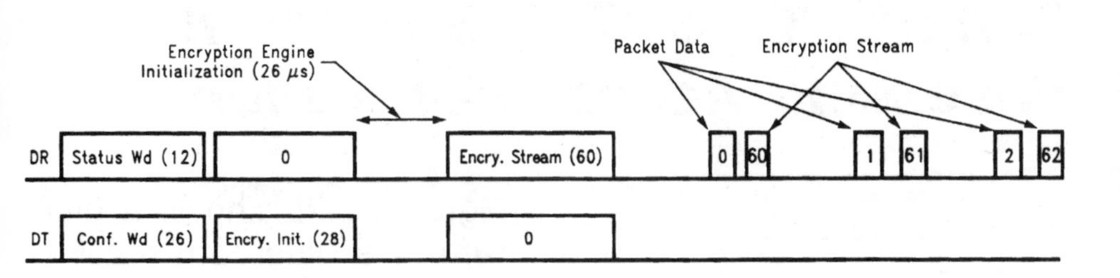

Figure 11.21. Receiving Bluetooth data

followed by a 60-byte encryption stream that allows the firmware enough time to prepare the encryption transmit data. Consequent encryption stream bytes piggyback with the packet data after the transition starts.

The Bluetooth profiles describe the following three security modes for Bluetooth devices.

- Security mode 1 (nonsecure). A device will not initiate any security procedure.
- Security mode 2 (service-level enforced security). A device does not initiate security procedures before channel establishment at L2CAP level. This mode allows different and flexible access policies for applications, especially running applications with different security requirements in parallel.
- Security mode 3 (link-level-enforced security). A device initiates security procedures before the link set-up at the LMP level is completed.

Bluetooth achieves authentication through pairing, or the entering of a personal identification number, but authorization occurs through the concept of trusted and untrusted devices. A trusted device has a fixed, or paired, relationship and has unrestricted access to all services. An untrusted device does not have a permanent fixed relationship but may have a temporary relationship and has restricted access. In some cases, a device may have a paired relationship but not have trusted status. Trusted devices receive automatic access, but other devices require manual authorization.

Philips PCF 87750 Bluetooth Baseband Access Controller

APPLICATIONS
AUDIO
TCP/IP
RFCOM
LOGICAL LINK CONTROL L2CAP
HOST CONTROLLER INTERFACE
LINK MANAGER
LINK CONTROLLER (BASEBAND)
RADIO INTERFACE
RADIO
ANTENNA

Figure 11.22. Diagram of a Bluetooth receiver

Figure 11.22 shows a block diagram of a Bluetooth receiver. Within the diagram, devices such as the Philips PCF 87750 Bluetooth Baseband Access Controller operate as a link controller and include the capability for encryption, decryption, authentication, and authorization. Referring to the block diagram shown in figure 11.23, the device integrates an embedded ARM7 TDMI microprocessor core, 384 kilobytes of Flash

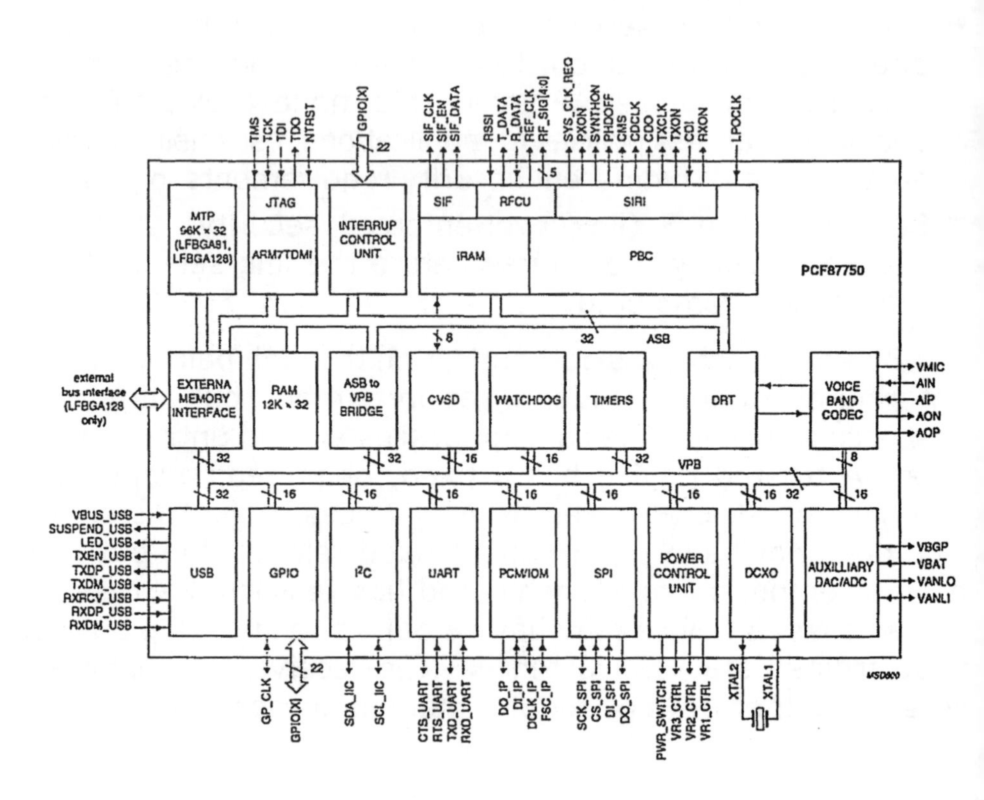

Figure 11.23. Block diagram of the Philips PCF87750 Bluetooth baseband controller

memory, 64 kilobytes of static random-access memory, voice band analog/digital converters, and a GFSK pulse shaper. The PCF87750 can run the complete Bluetooth application without requiring a host controller.

National Semiconductor LMX5001 Bluetooth Link Controller

Referring to the block diagram shown in figure 11.24, the LMX5001 Dedicated Bluetooth Link Controller operates as a master when communicating with the Bluetooth Link Management Controller/Host processor. The LMX5001 interfaces with the National Semiconductor Single Chip Radio Transceiver to provide a complete Bluetooth physical layer interface. During operation, the processor provides the LCI frame and data syn-

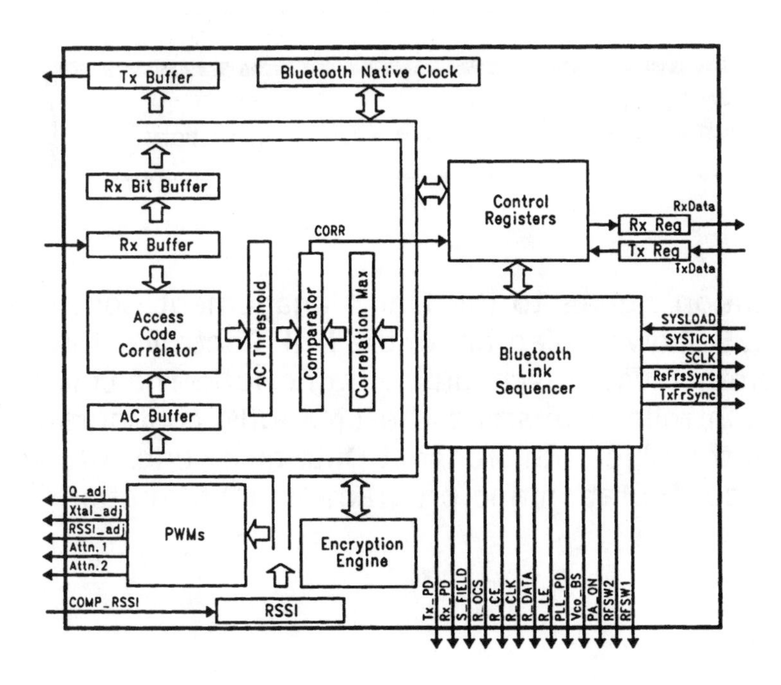

Figure 11.24. Functional diagram of the National Semiconductor LMX5001 Bluetooth link controller

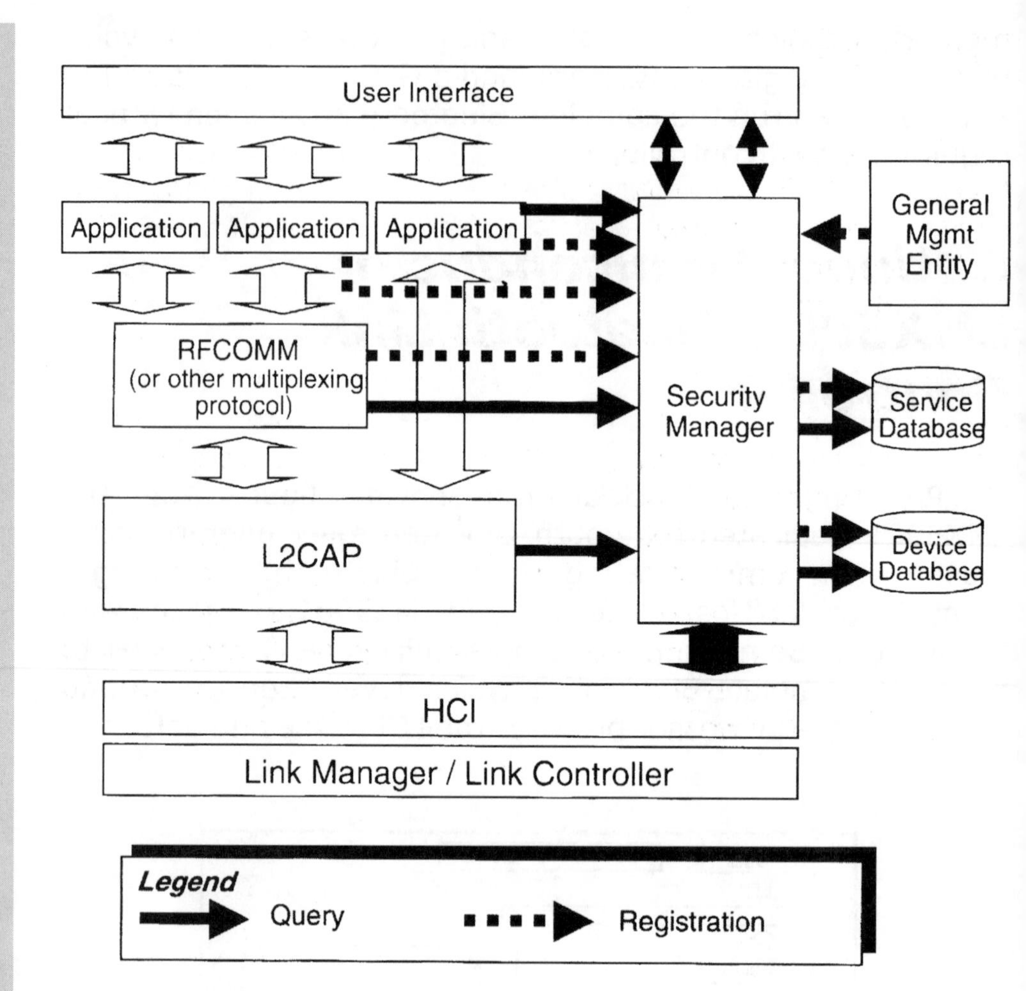

Figure 11.25. Diagram of the link-scheduling framework

chronization signals to the Link Management Controller. Figure 11.25 shows a diagram of the Link Controller Interface, or LCI, communication-scheduling framework. The Link Management Controller transmits the LMX5001 configuration data through the Bluetooth Transmit Data as a series of bytes that contain configuration and programming information.

HomeRF

The HomeRF group uses the Shared Wireless Access Protocol, or SWAP, standard to create a hybrid voice/data standard for the consumer market. HomeRF derived the SWAP technology from extensions of Digital Enhanced Cordless Telephone standard and the IEEE 802.11 standard for the purpose of producing a new class of wireless home services. The standard supports the application Time Division Multiple Access service to provide delivery of interactive voice and other time-critical services, and the Carrier Sense Multiple Access/ Collision Avoidance service for delivery of high- speed packet data. Based on the FHSS standard, the HomeRF network operates within the 2.4-MHz spectrum, offers a transmission power of 100 mW, and has a data transfer rate of 1 Mbps using 2FSK modulation and 2 Mbps using 4FSK modulation. Figure 11.26 shows the TDMA-CSMA/CA-TDMA sequence within the SWAP frame.

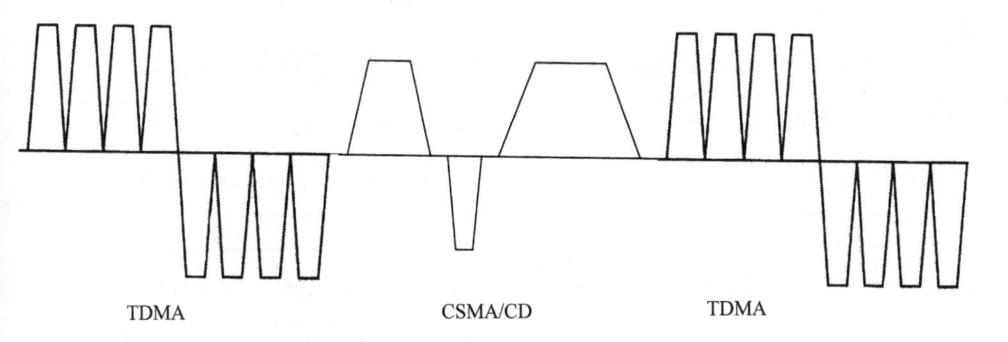

Figure 11.26. HomeRF SWAP TDMA-CSMA/CA-TDMA sequence

Figure 11.27 shows how the HomeRF system corresponds with the MAC layer. Within the data networking time, streaming media sessions have priority access. Moving to figure 11.28, the HomeRF Superframes and Subframes provide for all asynchronous and isochronous traffic. With the isochronous traffic, the addition of a beacon at the beginning of the frame reduces the frame length to 10 ms.

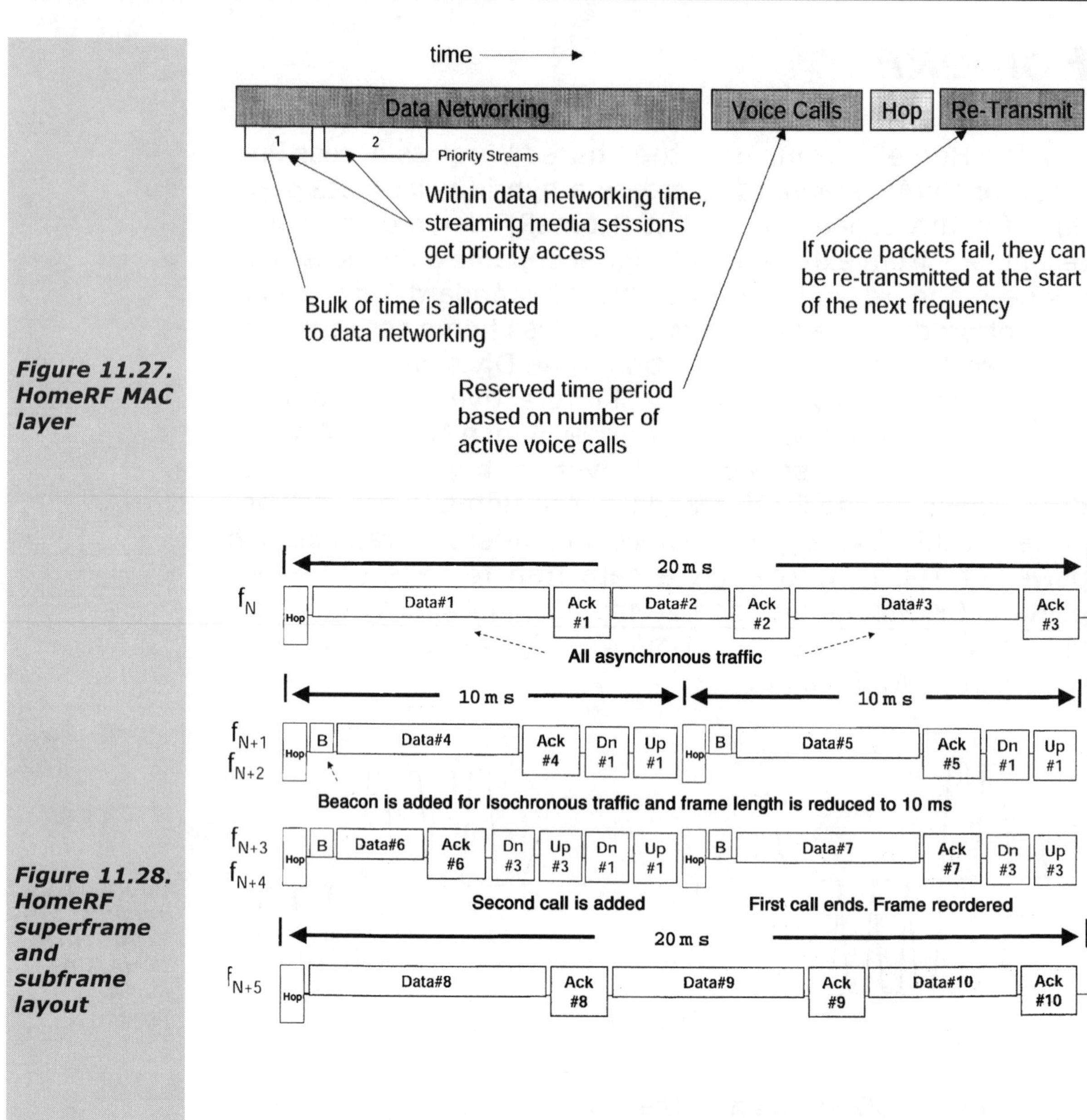

Figure 11.27. HomeRF MAC layer

Figure 11.28. HomeRF superframe and subframe layout

Through the use of the SWAP system and under the control of a Connection Point, HomeRF can operate either as an ad hoc network that supports only data communication or as a managed network that supports both data and voice communication. HomeRF offers a 10-Mbps peak data rate with fallback modes of 5, 1.6, and 0.8 Mbps. The wireless standard supports a simultaneous host/ client and peer/ peer connectionless topology and can support concurrent operation of multiple colocated networks through the use of a 48-bit Network ID code, or NWID. In the ad hoc network configuration, all stations have equal status with control of the network distributed between the stations. Connected through a standard PC interface, the Connection Point operates as a gateway to the Public Telephone Switched Network, supports power management through scheduled device wakeup and polling, and coordinates the system.

As figure 11.29 shows, a personal computer operates as the communication center for clients in the HomeRF system. The network can accommodate a maximum of 127 nodes including Connection Points, Voice Terminals that use only the TDMA service to communicate with a base station, Data Nodes that use the CSMA/CA service to communicate with a base

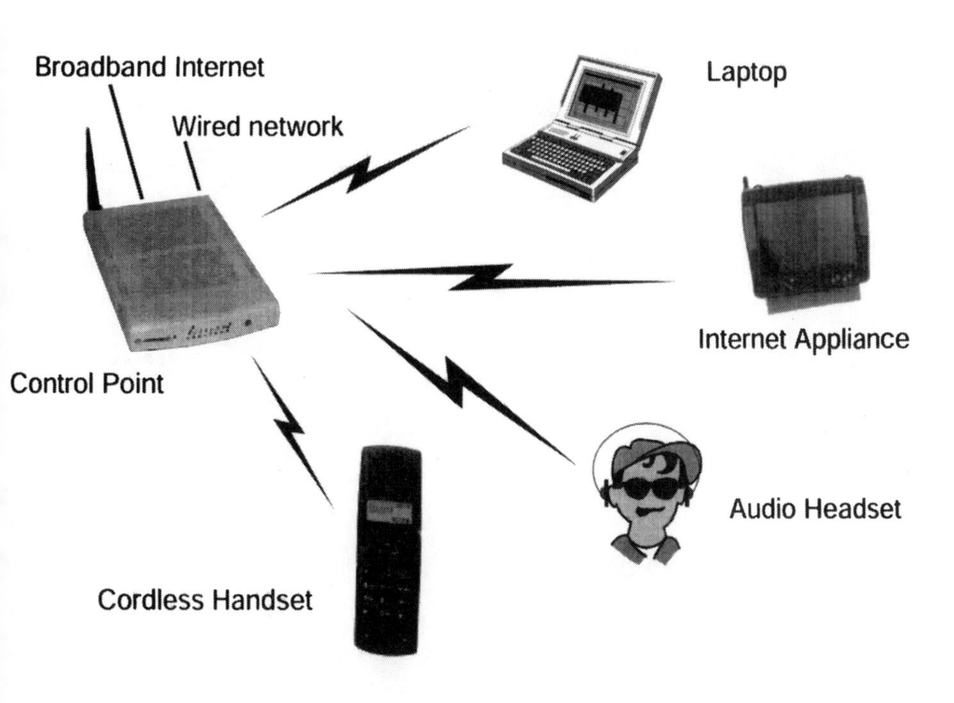

Figure 11.29. Illustration of HomeRF capabilities and operation

station and other data nodes, and Voice and Data Nodes that can use both types of services. In addition, HomeRF networks can support a maximum of eight simultaneous prioritized streaming media sessions for one-way or two-way sessions for audio and video transmissions. When configured in managed mode, HomeRF can also support a maximum of eight simultaneous toll-quality, two-way cordless voice connections with a maximum of six full-duplex conversations. The digital voice features include Call Line ID, Call Waiting, Call Forward, Intercom, and a suite of 911 services.

Along with integrating immunity to 2.4-GHz interference such as microwave ovens or other WLANs, HomeRF also provides several layers of network security based on the NWID. Figure 11.29 shows the method used by HomeRF to avoid active interference within the system. The system includes resistance to denial of service attacks through the random frequency hopping sequence for each NWID. HomeRF uses the Blowfish 128-bit data encryption and authentication algorithm.

12 Storage Area Networks

Introduction

Most major networking vendors agree that Storage Area Networks will move into a major role with distributed computing and data warehouse environments. Indeed, the vendors also agree that Storage Area Networks will become the accepted method for attaching and sharing storage devices to a network. Current uses for Storage Area Networks include connecting shared storage arrays, clustering servers in a server failover environment, taping resources to network servers and clients, and creating parallel and alternate data paths for high-performance or high-availability computer environments.

Storage Area Networks have begun to replace the direct-attached storage methods traditionally employed in networks. The benefits given through Storage Area Networks compare favorably with the higher costs and lack of scalability associated with direct-attached storage. Storage Area Networks provide scalability and high availability at high speeds.

As server clusters become more of a possibility for organizations engaged in eBusiness, eCommerce, and resource planning, the use of Storage Area Networks maintains the availability of the servers and addresses reliability issues. With the placement of data in a centralized storage location, the need for replicating data from server to server decreases. As a high-performance tool network management, Storage Area Networking safeguards an organization's critical data through storage and backup capabilities.

Defining Storage Area Network Operation

A Storage Area Network, or SAN, operates with the same type of hubs, routers, switches, and gateways that support local-area networks. However, the Storage Area Network pro-

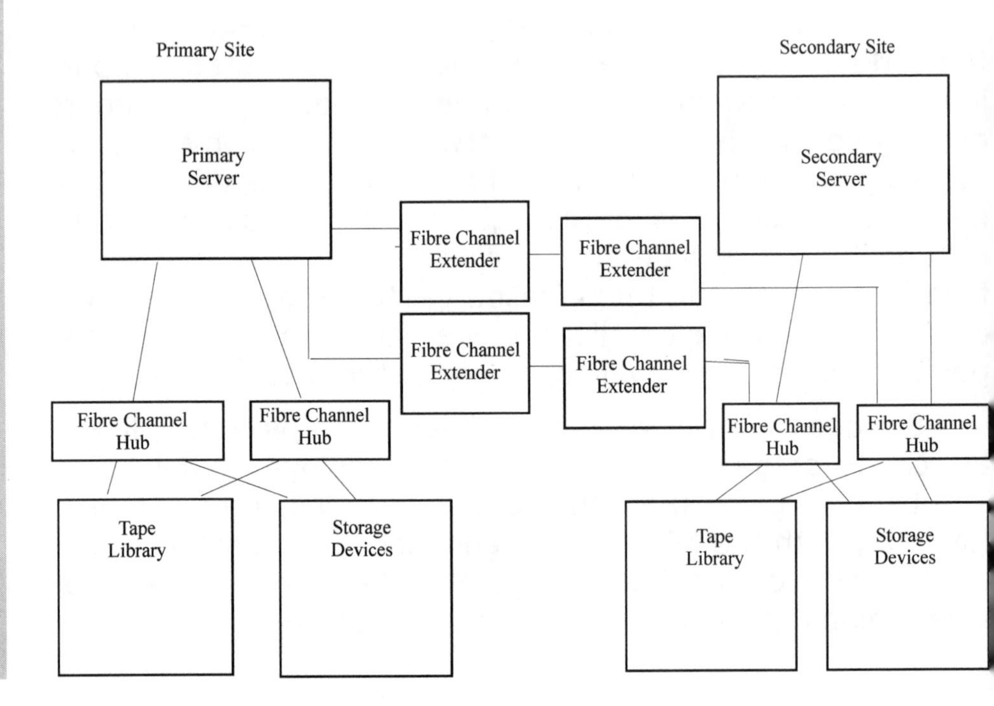

Figure 12.1. Mirrored SAN sites

Figure 12.2. A storage server

vides access to a high-speed network of shared, redundant storage devices. The network portion of the SAN establishes direct and indirect connections between multiple servers and multiple storage elements. As a result, the SAN extends the server storage bus across the network. The combination of hubs, switches, software, and interconnecting topology forms a fabric that serves as the storage bus.

As shown in figure 12.1, one type of SAN mirrors a primary site and a recovery site. The primary site includes a combination of servers, input/output channels, and storage devices. Software found at a server similar to the device shown in figure 12.2 provides backup, tape management, and hierarchical storage management capabilities. The input/output channels provide channel extension and conversion while connecting to a network that links the primary site and recovery sites. At the secondary site, the input/output channels connect to additional storage devices.

Fibre Channel

Fibre Channel provides an alternative to traditional serial interfaces and can operate with multiple protocols. Compared to parallel interfaces such as SCSI, Fibre Channel provides higher aggregate bandpass, greater connectivity of devices, and greater connection distances for Storage Area Networks. The scaleable Fibre Channel topology includes the point-to-point, switched fabric, and arbitrated loop connection meth-

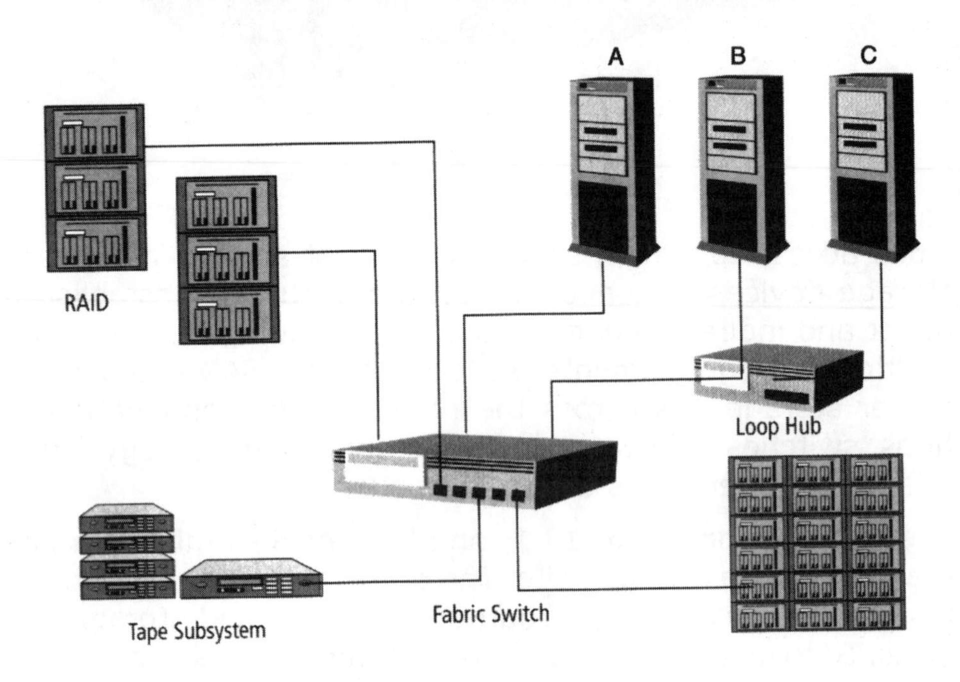

Figure 12.3. Fibre Channel-based Storage Area Network

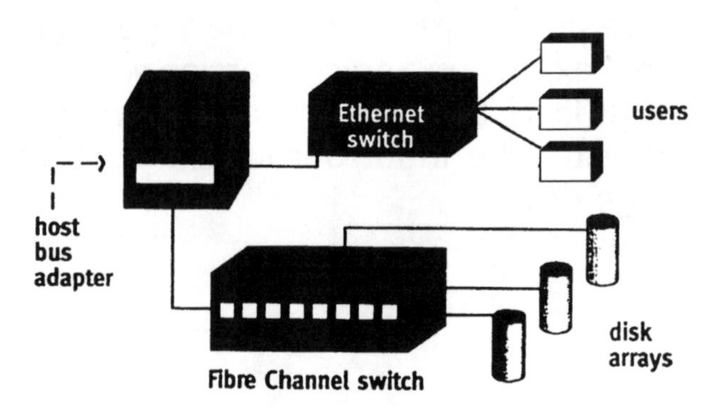

Figure 12.4. Use of a Fibre Channel switch in a SAN

ods. While figure 12.3 shows a SAN based on Fibre Channel connections, figure 12.4 depicts the attachment of a Fibre Channel switch into a SAN.

The use of Fibre Channel, or FC, supports 200-MBPS (megabytes per second) connectivity at a maximum of 10 kilometers from a source: however, the addition of optical link extenders substantially increases the connectivity given through Fibre Channel. Based on frames rather than packets, Fibre Channel can cluster a sequence of frames into a single block as large as 128 MB. During the transmission of data, the use of larger blocks decreases the number of interrupts. The use of a flexible circuit/packet switched topology combines with intrinsic flow control and acknowledgment capabilities to support connectionless traffic.

Fibre Channel Host Bus Adapters

Fibre Channel Host Bus Adapters, or HBAs, in the SAN offer the option to extend the storage bus to create a fabric that merges two Storage Area Networks into one large SAN. Using HBAs allows the addition of hosts and storage subsystems with no disruption to the loop. Every server in the fabric has access to every storage element in the fabric. Fibre Channel also supports increased network resiliency through dual loop capability. If a single loop becomes unavailable, the second loop continues to operate.

The transport rates of each storage device determine the deployment of disk arrays and individual disks in the Storage Area Network. A RAID combined with efficient Fibre Channel controllers can support a maximum 100-Mbps link speed. However, applications that do not require the maximum link speed may allow multiple RAID arrays to share a common loop segment.

Storage Area Network Topologies

As with other network technologies, Storage Area Network planning focuses on the implementation of specific network configurations. Each topology — point-to-point, star, switched fabric, and arbitrated loop — provides a solution to bandwidth requirements. In addition, each provides varying levels of support for the attachment of devices such as tape libraries.

The Point-to-Point Connection Topology

The point-to-point connection method for Fibre Channel involves the use of a single full-duplex connection between two devices. Because point-to-point connection does not include any intermediate devices, it provides the greatest possible bandwidth and lowest latency. However, point-to-point remains limited to two nodes.

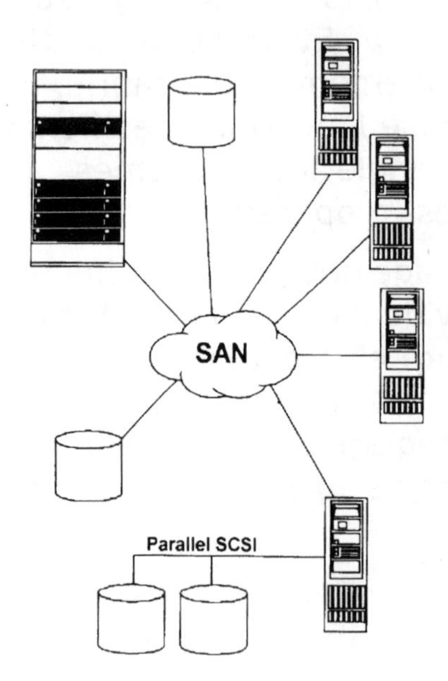

Figure 12.5. Star topology SAN

Star Topology

Shown in figure 12.5, the star SAN topology has gained widespread acceptance because the topology can support multiple points of destination in a localized cluster as well as centralized management. Although the star topology has the capability to share SAN-based systems, the topology does not provide large-scale scalability. However, the combination of the star topology and Fibre Channel can support medium-sized networks.

The Switched Fabric Connection Topology

When compared with the point-to-point and arbitrated loop topologies, the switched fabric method has the greatest connection capability and largest total aggregate throughput. Each device within the switched fabric connects to a switch and receives a nonblocking data path to any other connection on the switch. As a result, devices have dedicated connections to other devices on the network.

Figure 12.6 shows the switched fabric topology. When the number of devices increases to a point that involves the use of multiple switches, the switches connect together. Multiple connection paths between the switches increase redundancy and

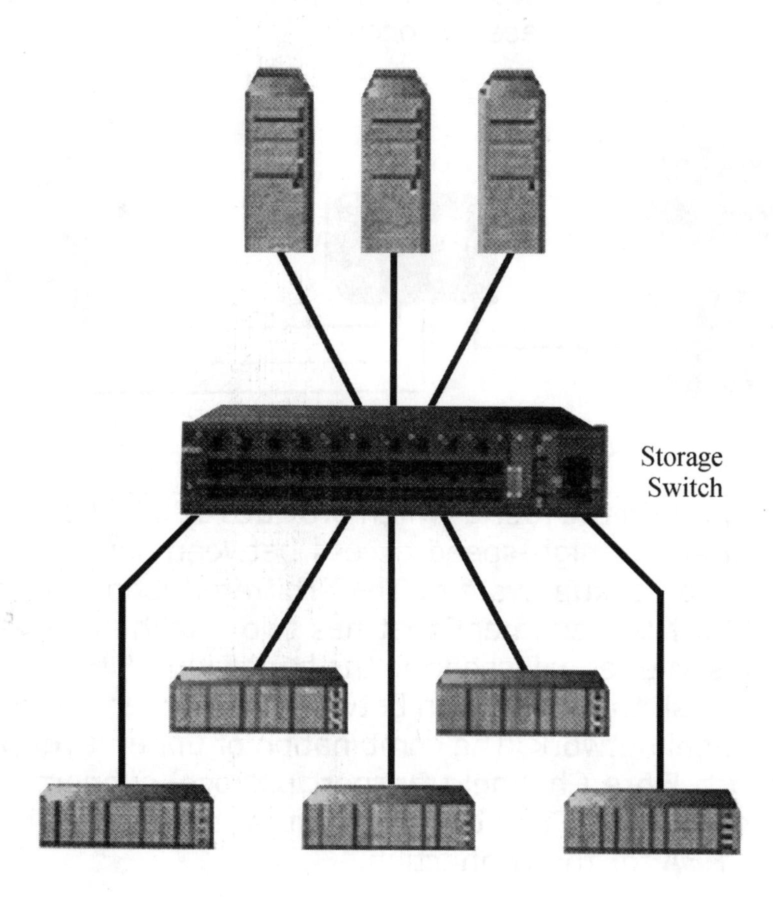

Figure 12.6. Diagram of a switch-based SAN

bandwidth. The switched fabric topology can interconnect large numbers of systems, maintain the high bandwidth requirements of a network, match the speed requirements between connections, and match cable differences.

The Arbitrated Loop Connection Topology

The arbitrated loop topology pictured in figure 12.7 distributes the switched fabric logic to all devices on the loop. As a result, each device can use the loop as a point-to-point connection. With the arbitrated loop connecting a maximum of 126 devices in a ring, all devices share the bandwidth of the loop. Each device contends, or arbitrates, for loop access. When the device receives access to the loop, a dedicated connection between sender and receiver occurs.

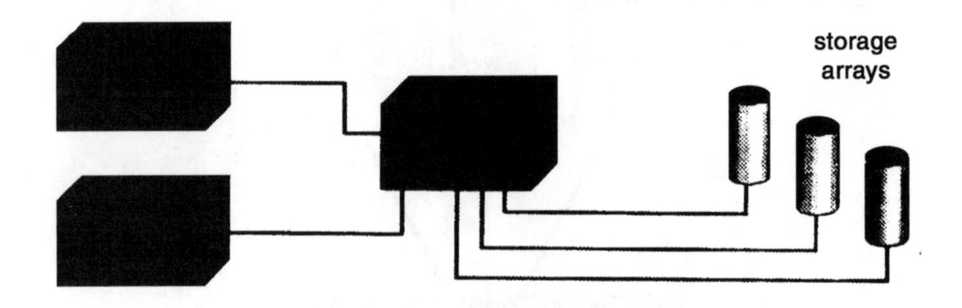

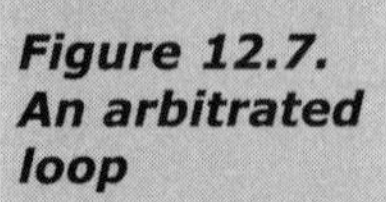

Figure 12.7. An arbitrated loop

Each port on the fabric switch provides 100 Mbps of bandwidth and allows high-speed access between servers, disks, and the tape backup system. The arbitrated loop hub represents a 100-Mbps segment that has two attached servers — with each server equally sharing the bandwidth. Fibre Channel HBAs establish the connection between the server bus and the Fibre Channel network. The combination of upper layer protocol support, Fibre Channel transport protocol support, bandwidth requirements, and bus architecture determines the appropriate HBA for the connection.

The arbitrated loop does not require a switch for the connection of multiple devices. As a result, the cost per connection substantially decreases. Because of the scalability and capability to sustain high bandwidth associated with arbitrated loop, the topology has become the most popular solution for Storage Area Networks.

Network Operating System Support for Storage Area Networks

Figure 12.8 shows a software flow chart for Storage Area Networks. Windows 2000, Novell NetWare, and Linux incorporate tools that enable more effective network storage architectures. Windows 2000 uses version 5.0 of NTFS, the NT File System, to increase the scalability of the NT network storage subsystem. NTFS version 5.0 provides the file and directory permissions needed to control access to the file system by network users and supports the high-capacity storage subsystems.

The latest version of NTFS support bulks up to 2 terabytes in size and also provides the capability to encrypt individual files and directories for added security. In addition, NTFS enables an application to create an extremely large file without expending disk space on meaningless data. As with Novell NetWare, NTFS 5.0 includes disk quotas, a feature that enables the automatic monitoring and restriction of the amount of disk space utilized by particular user. Furthermore, NTFS maintains a log called the Update Sequence Number, or USN, Journal that contains information about all file additions, modifications, and deletions on each volume.

In comparison to Windows 2000, version 5.0 of Novell Netware includes Novell Storage Service, or NSS. The use of

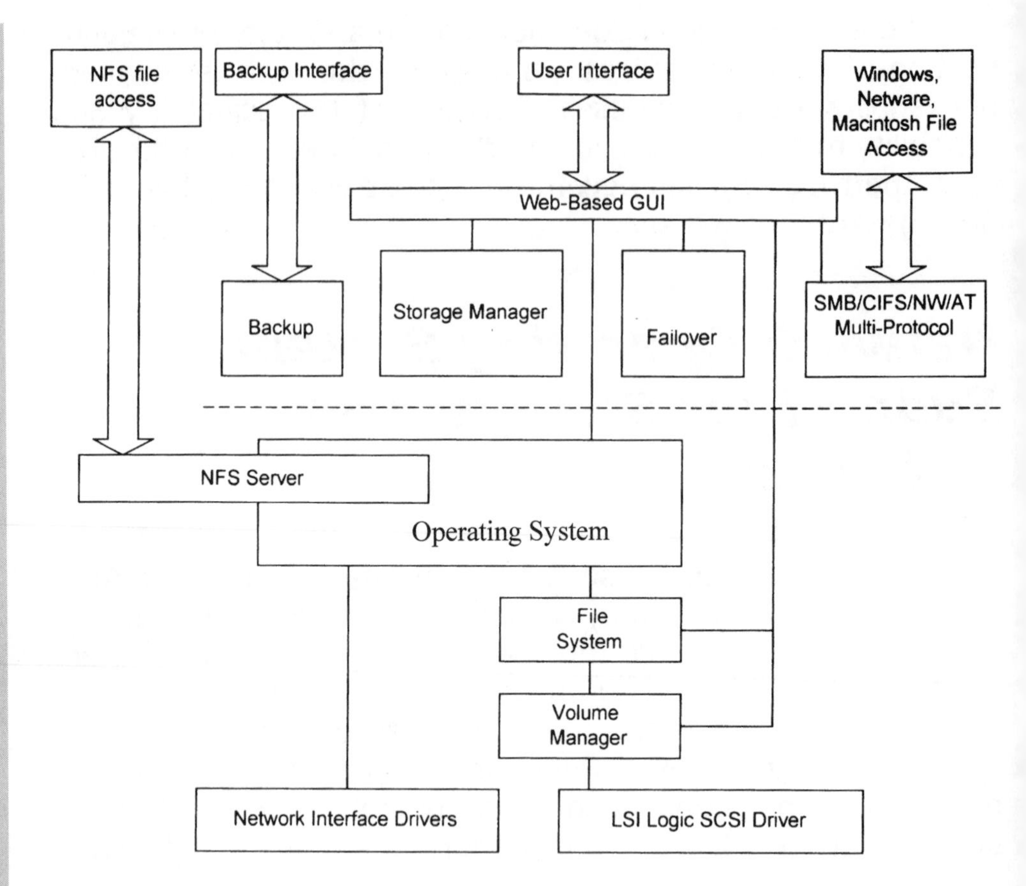

Figure 12.8. Block diagram of a SAN software system

the NSS 64-bit file system provides increased performance, supports a virtually unlimited amount of storage, and allows a NetWare server to host a maximum of 255 volumes with each containing a maximum of 8 trillion files. Individual files access through NSS can have a maximum size of 8 terabytes, and a user can open up to 1 million files at a time.

NSS combines the free disk space on multiple storage devices into a pool that allows the manual or automatic creation of NSS volumes. Novell Storage Service also mounts CD-ROM and FAT partitions as NSS volumes. In addition, NSS supports high storage availability through the capability to mount any size volume in less than a minute.

Various Linux releases support file sharing through compatibility with the Windows 2000 Server Message Block and the Novell NetWare Core Protocol. In addition, most — if not all — Linux variants include support for Network File System developed by Sun Microsystems. Supporting client-server applications, NFS allows a server to export all or part of a file system.

Local File Systems

In a SAN, the local file system controls the placement of user data onto locally attached storage devices without any need for network access. As figure 12.9 shows, a typical read or write request flows from the user to a disk. An application needing to perform a read or write passes a request to the kernel that indicates the name and type of the file, the location in memory for read data or the location for write data, and the number of bytes. Then, the file system converts the request into a read or write operation for a specific disk.

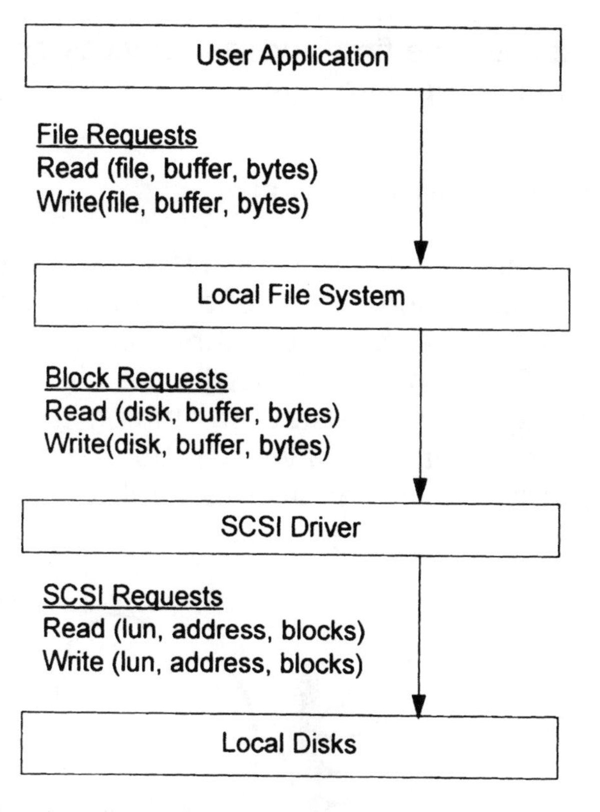

Figure 12.9. Local file system read-write sequence

The file system passes the request on to the SCSI driver where translation of the request into a SCSI command block understood by the disk occurs. Pronounced "scuzzy," the term

SCSI represents Small Computer System Interface and allows a variety of devices to be connected to a personal computer. With the connection made through a SCSI card installed within the computer, the SCSI standard offers the fastest available I/O connection, supports high-speed mass storage devices, and transfers data at rates ranging from 10 to 160 MBPS.

Using a unique identification number for each device connected to the SCSI chain, a SCSI controller card offers connectivity for internal and external peripherals and can connect up to 7 to 15 devices per channel. As a result, SCSI-connected devices can multitask and accept simultaneous read/write operations. The first and last devices on a SCSI bus terminate and stop the data signal.

Distributed File Systems

Referring to figure 12.10, note that a distributed file system moves some or all of the storage away from the client and onto a server by distributing access to data rather than storage. Files stored in a distributed file system remain accessible through any workstation running the distributed file system client software. If a workstation fails or becomes inaccessible for any reason, the user can utilize another client-capable sta-

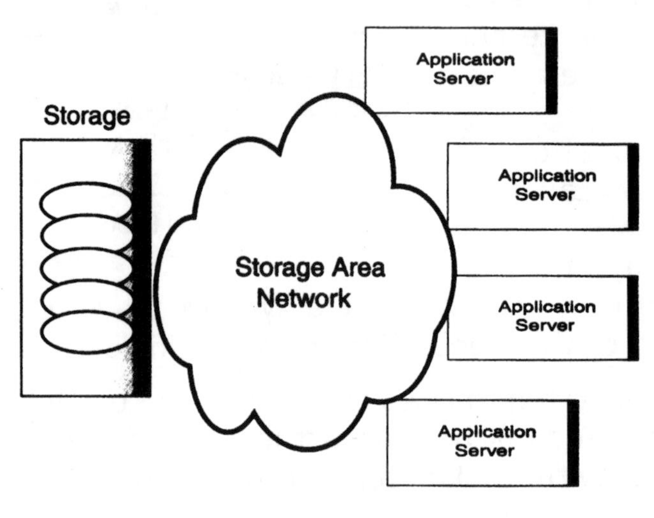

Figure 12.10. Distributed network-attached storage model

tion to run the same applications. The use of the distributed file system also promotes remote access; mobile users can have the same view of their data at a remote site as they have from their normal workstation.

To meet these needs, a distributed file system provides a common repository of storage available for partitioning among the users in increments smaller than an individual disk. As a method for increasing system performance and minimizing network traffic, distributed file systems allow the mixing of files between the local disk and the distributed file system. As an example, the file system allows data files to be moved to the centralized location for easy access and efficient backup.

During operation, the distributed file system adds a software layer called the redirector to the software at the client. As depicted in figure 12.11, the redirector determines the status of the file as either local or remote when the applica-

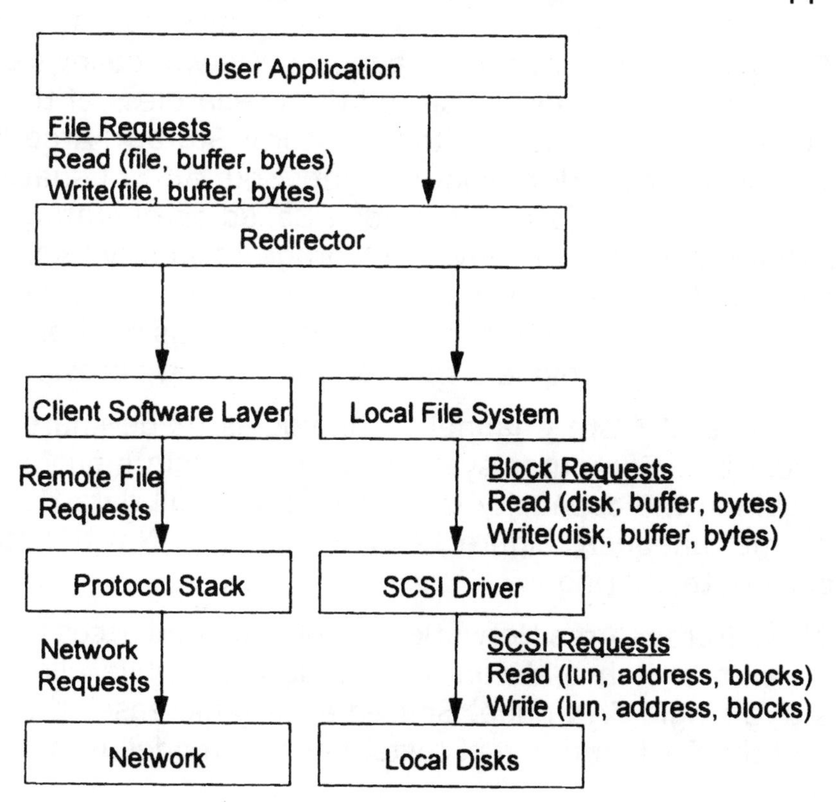

***Figure 12.11.* The redirector determines the status of the file**

tion requests the file. If the file has local status, the distributed system uses the same process as seen with the local file system.

If the file has remote status, the file system passes the request to the distributed file system client software. From that point, the client software determines which server owns the requested file and the protocol used to communicate requests to the file server. After the client software builds a request and passes the request to the protocol stack, the stack transfers the request to the server. Moving to the server, note that the received request is forwarded to the file server software for handling. The file server software interfaces to the local file system on the server to read or write the requested data.

Sharing Data Through a SAN

Sharing data through a Storage Area Network allows servers and clients to access all stored data regardless of the operating system or vendor. Although early Storage Area Networks shared network media, the networks did not share access to the data found on the media. Based on operating systems that did not handle physical storage, the early SAN solutions allowed the partitioning of the storage array across several workstations and the copying of data across multiple platforms but did not provide a method for sharing access.

The use of Fibre Channel technologies made shared access possible. Computer systems with multiple input/output ports gained the capability to share data at full data transfer rates simultaneously. Figure 12.12 depicts a SAN with heterogeneous data sharing.

Most Storage Area Networks rely on a shared access model that shares switch and controller resources while allocating disks to a single SAN node. Shared access increases the flexibility of the SAN for minimal costs. Another model uses shared

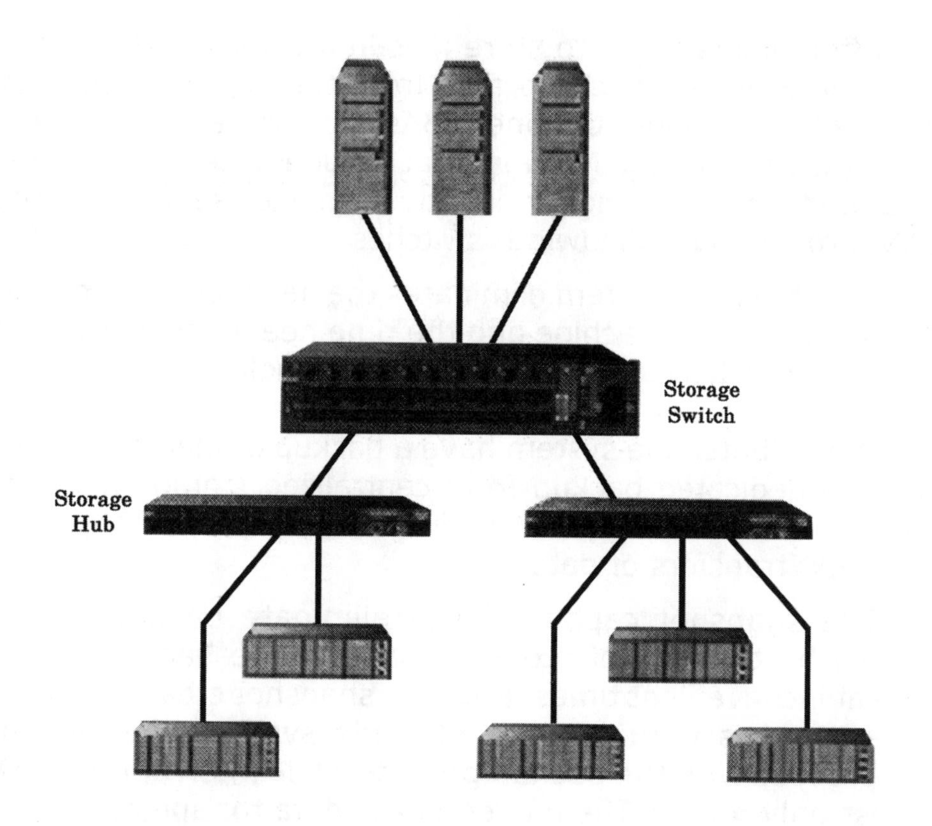

Figure 12.12. Diagram of a three-tier SAN

data clusters and a lock manager to synchronize file access. Each node acquires locks on disk blocks before attempting to access the block. The number of locks and lock-related messages increases with the number of cluster nodes.

Using the SAN for Backup

Because a SAN tape system attaches directly to the same storage as the servers, it can take advantage of an infrastructure that can offer data transfer rates of 100 MBPS or faster. Data backup occurs without any interaction with servers connected to the network and without any additional loading on the LAN. Rather than using LAN server processing power for the backup, the SAN uses processing based in the storage subsystem.

After clients begin to store files in a common file repository, the SAN can back up files from the shared repository directly to tape. Fibre Channel HBAs connect each server to a separate Storage Area Network that includes a centralized tape library. In turn, data moves directly from each server over the SAN through high-bandwidth switches.

With this, the system eliminates the need for backup utilities at each client machine and the time needed to set up and maintain the backup of each client. A single backup of the common file system ensures that all clients taking advantage of the distributed file system have a backup of their data. Providing a dedicated backup from centralized storage also improves network security. The dedicated backup eliminates bulk clear text transfers of data.

The snapshot capability can eliminate backup issues caused by the lack of storage space or the backing up of data at inconvenient times. Because snapshot creates a near-instantaneous virtual image of a file system, the system can proceed with the backup process while interrupting data access only briefly. The movement of data to tape can occur while data access continues. Network administrators may remotely share snapshot images for read-only access by applications or for the purpose of recovering a deleted or corrupted file.

Synchronous and Asynchronous Storage

With synchronous storage, the local system mirrors data to the remote site as an integral part of input/output command processing. If the primary site or system becomes inoperative, the secondary or remote copy can keep the system running after the users have switched to the secondary site. Even though synchronous storage offers a common approach to mirroring, it requires high bandwidth and remains limited in terms of distance between networks. As the name implies, asynchronous storage places local I/O writes into a queue for later transmission. Although asynchronous storage offers a risk for losing buffered or in-progress data if the primary system fails, it provides better bandwidth and distance performance.

Disk-Based Storage

Along with utilizing automated tape storage systems, Storage Area Networks also rely on the sharing of disk-based storage among multiple servers. The network uses ports on the switch to divide a common disk array into separate volumes. SAN file-sharing systems allow multiple users to share common data down to the individual file level and support storage-intensive applications such as video production. The access to centralized files at local disk speeds occurs through the SAN bandwidth and decreases disk requirements, eliminates the need for duplication, and cuts the need for transferring files.

Intelligent Tape Libraries

Modern Storage Area Networks require intelligent subsystems that can manage bandwidth, access control, and compatibility with third-party systems. Intelligent tape libraries integrate a PCI expansion bus that supports the installation of PCI server cards. With the installation of the server card, the tape library operates as an intelligent SAN appliance and provides library sharing as well as network management. The tape library can automate backup functions, provide centralized control, allow remote administration, and provide diagnostics.

The use of intelligent systems also allows the transparent selection of the location of particular files by users and systems administrators. Placement of the data within the center of the enterprise removes the linkage of data to specific locations. With this, the selection will depend on the type of data and demands for use of the data.

Fault Tolerance

To improve fault tolerance, the Storage Area Network incorporates RAID disk and tape technology for both improved reliability and performance. Additional fault tolerance in a SAN occurs through the configuring of Fibre Channels and the switched fabric technology in dual-redundant modes. The configuration ensures that no single point-of-failure within the Storage Area Network can occur and allows the swapping of failed Fibre Channel components without shutting down the system.

Remote mirroring through a SAN can provide an off-site copy of business critical data, data that can be immediately accessed in the event of a catastrophic loss of the data at the original site. For each application write, remote mirroring software issues a write I/O operation to the local disk pool and a second write I/O across a communication link to the remote site. With the data safely protected off-site and still accessible by users, a business that has experienced a major system failure can quickly move past the costly and time-consuming data recovery process.

Fault tolerance also exists through the capability of RAID arrays to map physical disks to logical volumes protected against disk failures. The arrays can export the volumes to servers across shared devices. Controllers used with the arrays provide storage management through volume mapping, remote mirroring, and the use of virtual volumes.

With the application of virtual volumes, the controllers support the aggregation and subdivision of disks to SAN nodes and the incrementalization of usable capacity to the storage pool. Looking at figure 12.13, note that controllers create virtual copies of on-line data in a temporary storage area and save the "before state" of on-disk data. Then, the data structure points the backup operation to the on-line volume for blocks that have not updated after the creation of the virtual copy.

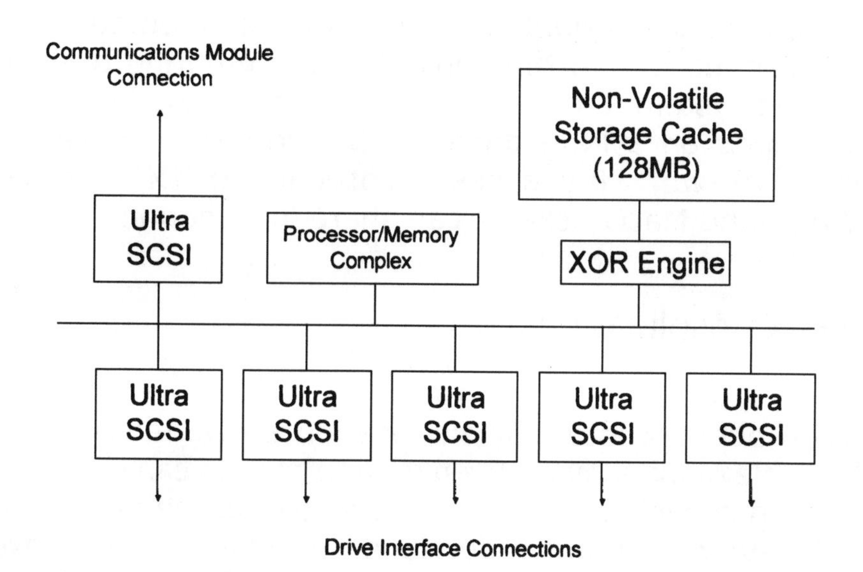

Figure 12.13. Block diagram of the RAID controller architecture

Storage Area Network Benefits

Most user applications — such as electronic mail, FTP, communication across and downloads from the Internet, and eBusiness — require the transfer of data across a network. A local-area network provides an infrastructure for moving files from location to location within a small department, but a SAN can provide an alternate means of file transfer and free the LAN for other communication. Because the SAN encompasses all storage devices, movement of data from disk to disk, disk to tape, or disk to optical occurs independent of the local-area network. As a result, the SAN becomes a working complement to LANs and WANs.

Reduced Demands for Local Storage

Any authorized user can access files stored in a distributed file system. As a result, workgroups that rely on consistent information sharing can easily retrieve updated files. Configuration of the distributed file system can automatically establish connections to the network storage during the initiation of the system. From a user's perspective, the files appear to be stored on a local drive. If the file server implements multiple network file access protocols, UNIX, Windows, NetWare, and MacOS clients can share information.

Server Consolidation

With figure 12.14, the consolidation of file servers stretched across a network into a single location allows more efficient and reliable management. Even though servers can share high-performance backup hardware and software, the network can also achieve full storage redundancy. In addition, the network can have more flexibility with the allocation of storage and can allow for variations in storage demands.

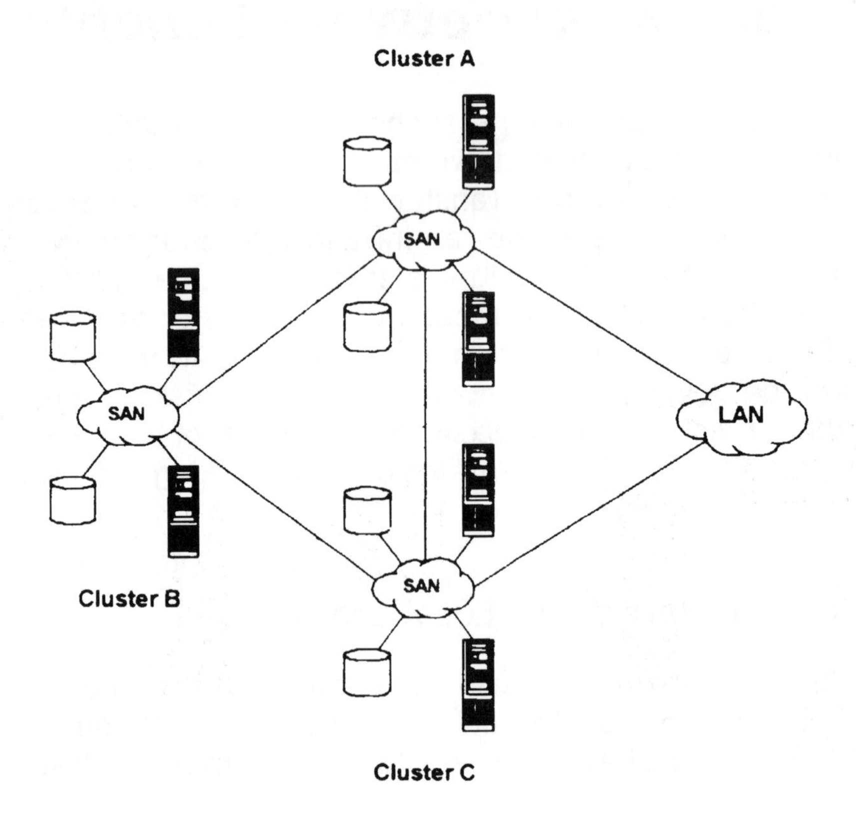

Figure 12.14. Clustered servers

Network Attached Storage

Network Attached Storage can replace the RAID storage subsystem or tape silo of a Storage Area Network and provides a server-to-server storage solution. As pictured in figure 12.15, a Network Attached Storage, or NAS, uses the local-area network interface and communications protocol to connect servers to servers. In contrast to a SAN that does not run a network operating system, network-attached storage products implement one or more distributed file system protocols that allow both clients and servers to access files stored in a common, shared storage pool. Consisting of a group of hard disks equipped with a processor and embedded operating systems, NAS servers support cross-platform file sharing and environments that mix Windows, Apple Macintosh, and UNIX/Linux workstations.

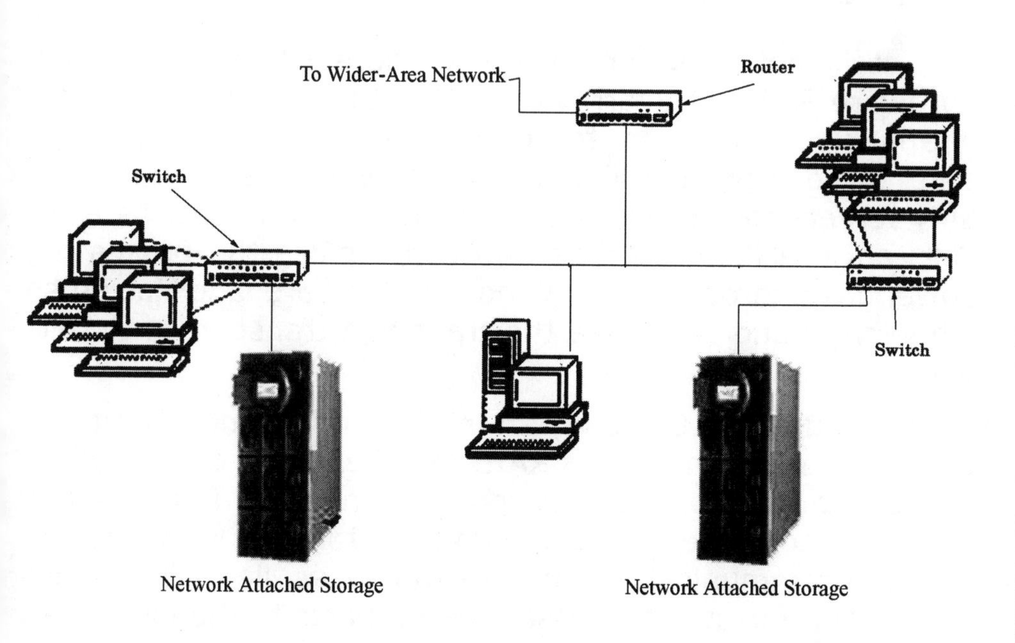

Figure 12.15. Diagram of network-attached storage

The storage of files in a common repository allows

- the sharing of files among users,
- the access to files by the same user from different locations,
- the easy and economical addition of storage,
- the partitioning of storage for users,
- the efficient backup of files, and
- the consolidation of multiple file servers into a single managed storage pool.

With Network Attached Storage, one server operates as a direct attachment to the storage devices. All attached servers maintain large storage requirements by passing data to the NAS server over the LAN. Because NAS disk arrays organize around an embedded operating system established for storage operations, an NAS can install directly into the corporate network. The devices take advantage of Ethernet networking attributes and remain open to support by SNMP.

Intelligent Hubs

Intelligent hubs combine one or more network interfaces, hardware RAID disk arrays, an AIT tape library, and hot-swappable components to provide a complete storage solution. In addition, intelligent hubs support the migration of LAN servers to a consolidated server environment. As a result, an intelligent storage hub can replace the processing power given through UNIX, Windows, NetWare, and Macintosh file servers and can make the transition transparent to network clients.

As figure 12.16 shows, an intelligent hub incorporates a communications module, a RAID module, and a tape module. The communications module provides network interfaces to the LAN and processes all network protocols. During operation, the communications module translates file requests into block I/O requests and passes the requests to the disk array

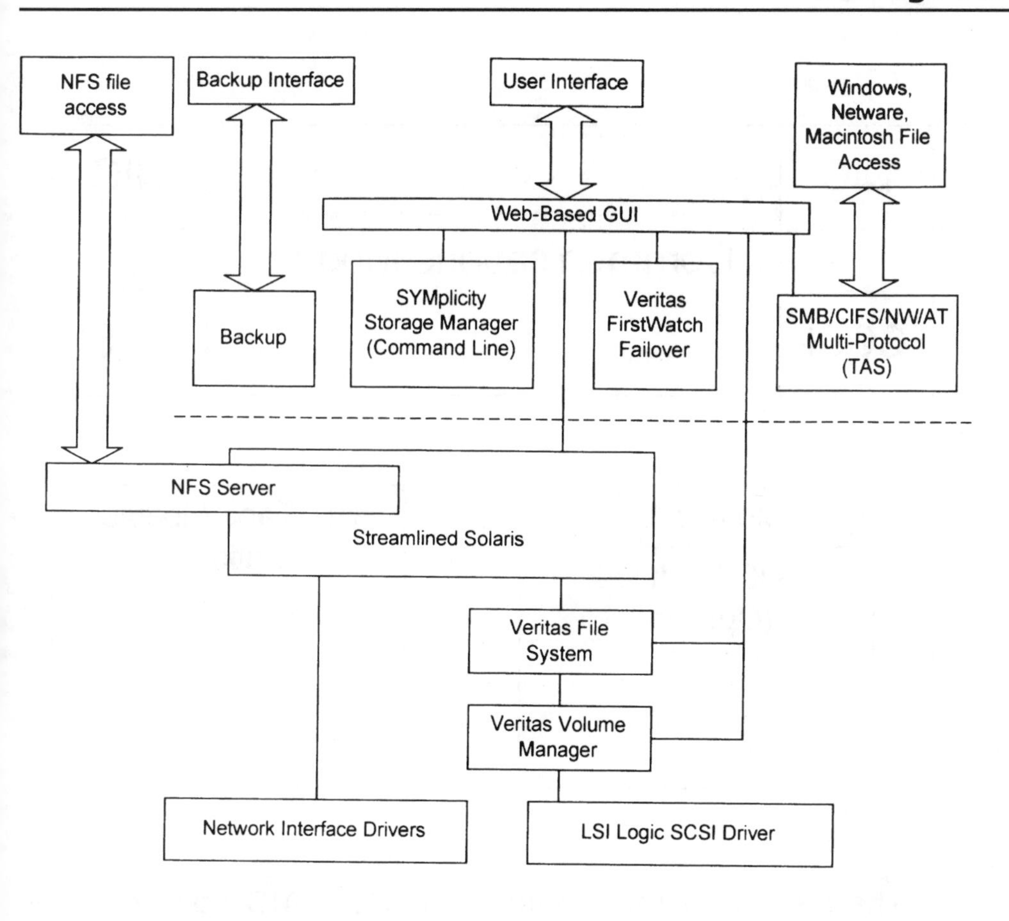

Figure 12.16. Block diagram of an intelligent hub

module. Although the communications module can operate without a dedicated console, an RS-232 serial port allows optional maintenance access.

Looking at figure 12.17, note that the communications module provides the internal bandwidth and processing capability needed to move data between the network interfaces and the RAID module. As the accesses to the NAS occur, the communications module services network requests data or file information from its cache and does not require disk array access. Because no write caching occurs on the communications module, the device ensures that any hardware, software, or power failures in the communications module cannot cause data loss.

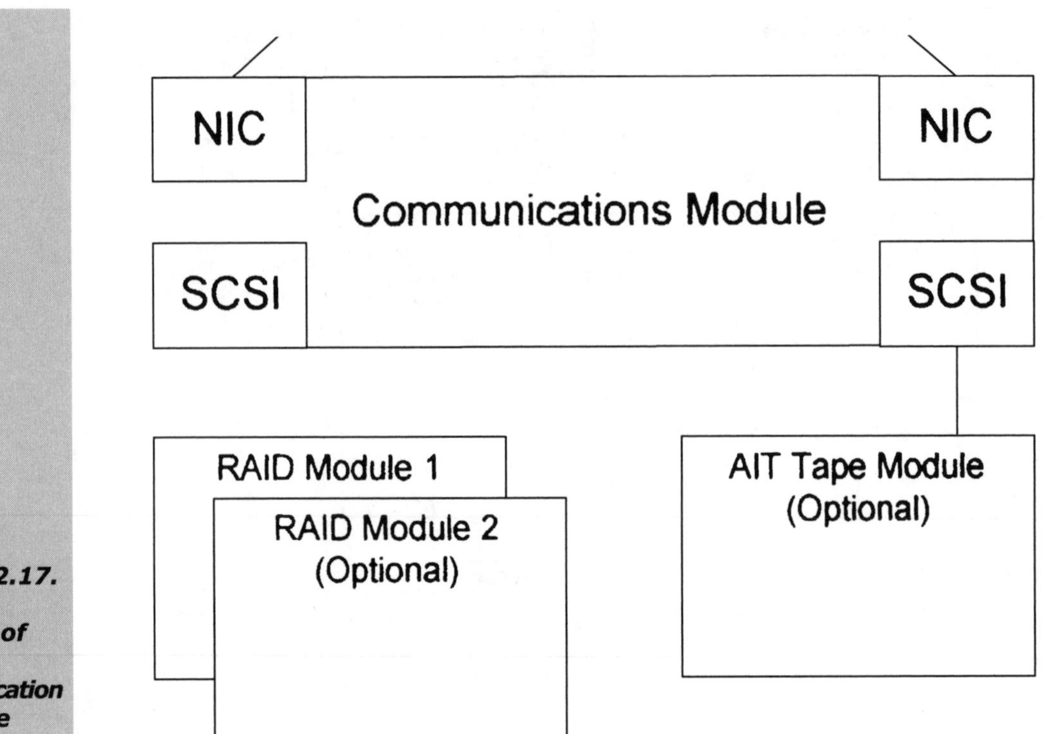

Figure 12.17. Block diagram of SAN communication hardware

The disk array module maintains all RAID operations on the data and provides the storage foundation for the intelligent hub. In addition, the RAID module software controls all array data placement, check data management, and array failure recovery. From the vantage point of the communications module, each array logical unit appears as a single large disk. Ultra SCSI interconnections establish 40-Mbps data paths between the RAID module and the communications module.

The disk array controller allows the off-loading of all aspects of the RAID data protection algorithms from the communications module onto a dedicated, embedded controller. Specialized, high-speed hardware on the controller performs RAID parity calculations while the system operates. Independent backend SCSI buses allow the transfer of data in parallel to the disk drives that comprise the redundant array.

To ensure the safety of critical data, the disk array module includes redundant and hot-swappable disk drives, fans, and power supplies. The dual controller design protects against the loss of a controller, drive channel protection, and write cache mirroring. Using software control, the module monitors the I/O status of the controllers. If an error occurs and indicates that any part of the I/O path from the host adapter in the communications module through the RAID controller has failed, the software dynamically reroutes I/O operations through the other RAID controller.

The drive channels play an important role within the RAID module architecture. If one channel fails, the system maintains optimal availability by showing the array with a failed drive in each drive group and then using standard RAID recovery algorithms. The system also mirrors write cache data on the controllers before acknowledging that a write sequence has occurred.

Using drive channels with a 200-MBPS data transfer rate between caches minimizes the impact on I/O operations. The Processor/Memory complex interprets the command, prepares the appropriate drive commands, and uses one or more of the intelligent Ultra SCSI processors on the drive interface and communications module interface to transfer the request and data to and from the drives. The Nonvolatile Storage Cache supports read and write caching and uses advanced algorithms for cache management.

An optional tape module stands as the final architectural piece of the storage hub. The module consists of one to four Advanced Intelligent Tape, or AIT, tape drives housed in a tape library that has a capacity of 20 or 40 cartridges. Under control of the communications module, backups and restores can execute on demand or become scheduled for automatic execution.

13

VPN and VLAN Technologies

Introduction

This chapter looks at two different network applications that use the connectivity given through wide-area networks. Virtual Private Networks, or VPNs, facilitate access from any location reached by the Internet. Given the worldwide presence of the Internet, a VPN could stretch across international borders and reach most major cities. In addition to extending the reach of a corporate network, Virtual Private Networks create a secure communications ring.

Virtual Local-Area Networks, or VLANs, also extend the reach of the enterprise by creating of the look and feel of a LAN over a wide-area network. Rather than rely on the traditional network segmentation seen with local-area networks, VLANs provide segmentation according to the purpose of a workgroup. To accomplish this type of segmentation, VLANs rely on switching technologies. As with Virtual Private Networks, VLANs can provide enhanced security for an organization.

Defining Virtual Private Networks

A Virtual Private Network uses the Internet or other network service as its WAN backbone. Local connections to an Internet Service Provider or the point of presence of other service providers replace dial-up connections to remote users and leased-line or Frame Relay connections to remote sites. With this, a VPN extends a private Intranet across the Internet or other network service and facilitates secure eCommerce and Extranet connections between business partners, suppliers, and customers.

A Virtual Private Network operating within a company and with resources managed by a single organization establishes an Intranet. When VPNs connect two or more networks between corporations, the combination forms an Extranet. With an Extranet VPN, no single organization exerts management control over all network and VPN resources. Instead, each company manages its own VPN equipment. Implementing an Extranet VPN requires configuration of a portion of the VPN and the exchanging of subsets of the configuration information with partner VPN management organizations.

The deployment of a Virtual Private Network should involve careful planning and an appraisal of network needs and infrastructure. Planning processes should identify VPN applications, determine bandwidth requirements, and align the implementation of the VPN with current network security policies. The processes should also consider network management issues such as public and private addressing schemes, translations, naming services, and placement of VPN devices relative to existing routers, firewalls, servers, and bandwidth management devices.

VPNs and Intranets

Sometimes referred to as site-to-site or LAN-to-LAN VPNs, Intranet VPNs extend and secure private networks across the Internet or other public network service. Compared to other methods such as leased-line solutions, the implementation of Intranet VPNs can provide cost-effective branch office networking and offer significant cost savings. The Intranet site categories include small office/home office (SOHO) sites, branch sites, central, and enterprise sites. As pictured in figure 13.1, an Intranet VPN utilizes either local ISP connections to the Internet, secure Frame Relay, or secure ATM connections rather than long-distance leased lines. In addition, last mile technologies such as Digital Subscriber Line (DSL), networks, cable networks, or wireless networks provide high-speed Internet access for the Intranet VPN.

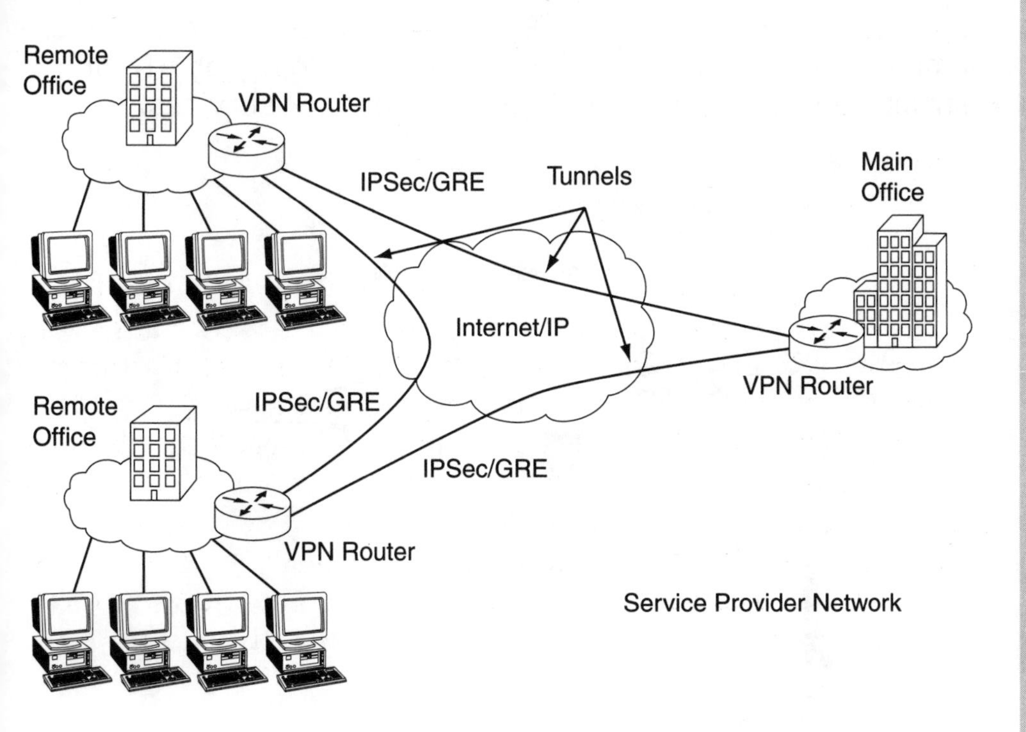

Figure 13.1. An Intranet VPN

Intranet VPN benefits include lower costs for line rental, scalability, lower costs for backup, lower costs for bandwidth over the last mile, and lower costs for backbone services. Typically, VPN carriers provide a leased-line feed by contracting with a traditional carrier company. Because leased lines sometimes have a distance-related cost structure, connecting to a local point-of-presence will provide savings compared to a direct long-distance or international link.

VPNs and Extranets

As depicted in figure 13.2, Extranet VPNs operate as an extension of Intranet VPNs and allow secure eCommerce connections between business partners, suppliers, and customers through the use of firewalls. Specific network types — such as ISDN, Frame Relay, or ATM — provide different benefits because connections for those networks have different operating characteristics than VPNs based on IP tunnels. For example, a VPN based on an ISDN, Frame Relay, or ATM infrastructure uses public switched data network services.

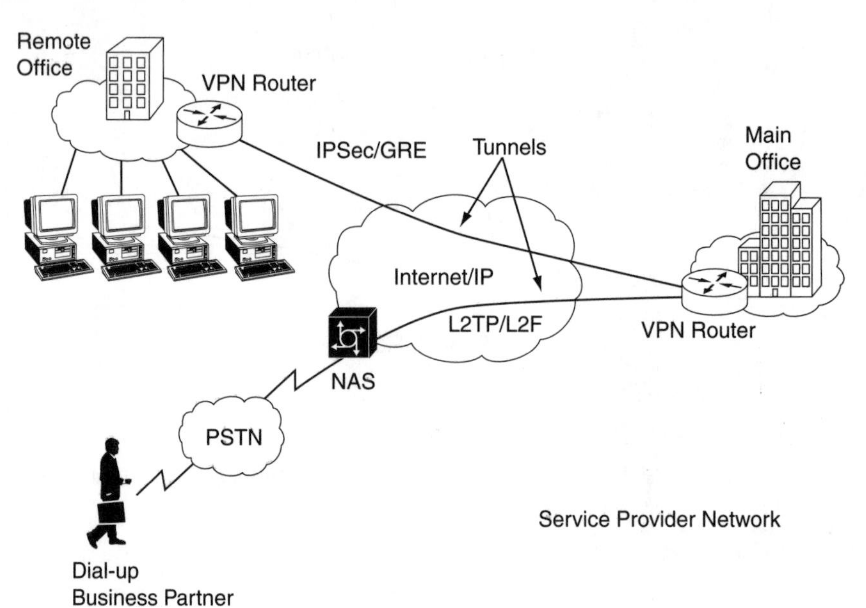

Figure 13.2. An Extranet VPN

ISDN

An Integrated Services Digital Network, or ISDN, provides an on-demand digital telephony and data-transport service between two points. In brief, the ISDN technology digitizes the telephone network and permits the transmission of voice, data, text, graphic images, music, video images, and other information over existing telephone connections to devices attached to local-area networks.

We can categorize ISDN services as bearer services that use lower layer transport facilities or as teleservices that have distributed application capabilities. Bearer services allow a user to transfer information to another user without restriction on the type or format of the data. ISDN teleservices include telephony, facsimile, interactive videotext, and videophone and offer guaranteed end-to-end compatibility regardless of the terminal equipment.

Frame Relay

Frame Relay provides a high-performance, packet-switched WAN protocol that operates at the Physical and Data Link Layers of the OSI reference model. Originally designed for use across ISDN interfaces, Frame Relay has become useful over a variety of other network interfaces. Frame Relay services may interconnect with other types of network services and provide high-bandwidth connectivity over a variety of interfaces.

ATM

ATM networks use cell-switching and multiplexing technologies that combine the benefits of circuit switching and packet switching. Originally intended for public networks, the connection-oriented ATM has gained usage in both public and private networks. With cell relay, the network conveys voice, video, or data information in small, fixed-size cells. The use of circuit switching in an ATM network provides guaranteed capacity and transmission delay, while the use of packet switching provides flexibility and efficiency for intermittent traffic. As a whole, ATM networks offer scalable bandwidth that ranges from a few megabits per second to many gigabits per second.

The connections support any type of communication including video conferencing, private data communications, and international communications. In addition, ISDN, Frame Relay, and ATM providers have established extensive billing and accounting information bases. Because data travels over a private network owned by a service provider or carrier, security becomes less of an issue for VPNs based on ISDN, Frame Relay, or ATM networks.

VPNs based on the Internet have become a widely available alternative to dial-up remote access or to a private network based on public network services such as T1 leased lines or Frame Relay. Because of the widespread use of the Internet, organizations have begun to apply Internet VPN technologies for internal communications. In addition, Internet-based VPNs also provide a flexible method for outsourcing remote access at significant cost savings. However, Internet VPNs may not offer adequate levels of security, quality of service, data throughput, or latency guarantees.

VPN networks combine all traffic over the connection used by the router for Internet access. As a result, the implementation of the VPN does not include a Remote Access Server (RAS), modem banks, and ISDN terminal adapter pools. With the simplification of equipment needs, network staff assigned to the VPN only manage a high-performance router. With outsourced VPN services, the ISP or carrier can manage the router and produce additional cost reductions.

Public networks relying on ISDN, Frame Relay, and ATM connections can carry mixed data types and can provide VPN services through B channels, Permanent Virtual Circuits (PVCs), or Switched Virtual Circuits (SVCs) to separate traffic from other users. Using B channels, PVCs, or SVCs provides additional bandwidth and quality-of-service capabilities based on usage. As a result, applying the services for VPNs also offers cost savings.

Establishing a VPN over a Frame Relay network moves away from expensive dedicated leased lines and takes advantage of Frame Relay benefits such as bandwidth on demand, support for variable data rates for bursty traffic, and switched as well as permanent virtual circuits for any-to-any connectivity on a per-call basis. Because of the optimum use of bandwidth, Frame Relay presents better latency and performance for the VPN.

Implementation of a VPN through Frame Relay occurs through the creation of a mesh of Frame Relay connections between sites or the use of IP tunnels over Frame Relay connections between sites. The Frame Relay connections between sites operate as point-to-point links and have characteristics similar to dedicated leased lines. Data remain separate from other Frame Relay users because each connection uses a separate virtual circuit.

With the use of IP tunnels over Frame Relay, the connections between sites again resemble the point-to-point links seen with dedicated leased lines. In addition, each connection uses a separate virtual circuit. In comparison, though, several separate IP tunnels can run over each connection. Each tunnel will accept encryption and authentication for provide additional security.

VPNs based on ISDN, Frame Relay, or ATM services have disadvantages as well as advantages. Because the technologies have not gained widespread usage, costs may remain higher. In addition, the broadband services may not provide total support for Extranet and eCommerce connections to business partners, suppliers, and customers.

VPNs and Remote Access

In previous chapters, we defined remote access as the ability to connect to a network from a distant location. A remote access client system connects to a network access device such as a network server or access concentrator. Sometimes called dial VPNs, Remote Access VPNs allow individual dial-up users to connect securely to a central site across the Internet or other public network service. When logged in, the client system becomes a host on the network. Typical remote access clients include laptop computers equipped with modems, personal computers that rely on modems or other types of connectivity for connections from a home or office, and laptop computers connected as part of a shared local-area network.

Remote access connections divide into either local dial or long-distance dial groups. For the private remote access networks shown in figure 13.3, local-area users connect through different types of telecommunication services. Remote access long-distance users typically rely on modem access over telephone networks and gain dedicated switched access through channelized leased-line and primary-rate ISDNs. The VPN shown in figure 13.4 relies on dial-in access; however, access to the VPN shown in figure 13.5 occurs through both dial-in facilities and telecommunication services.

Given the enhanced flexibility and reduced costs available through these technologies, Remote Access VPNs have begun to replace traditional remote access solutions. With VPNs, local-area users can select from a wider range of data services regardless of the support given at either the enterprise or central site VPN equipment.

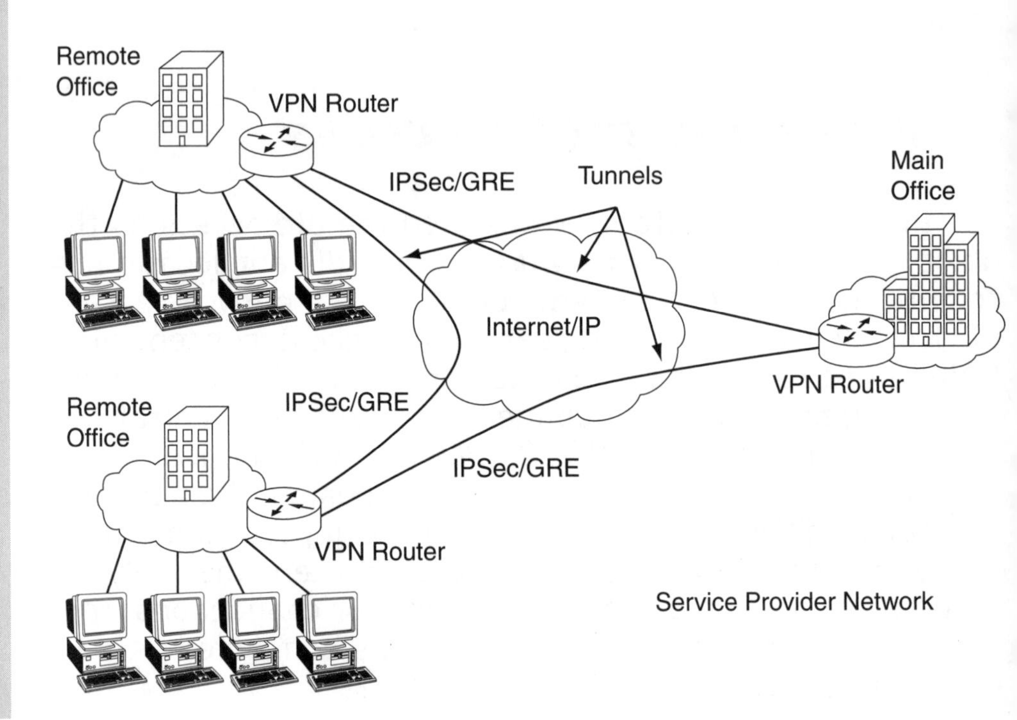

Figure 13.3. Remote access to a Virtual Private Network through broadband technologies

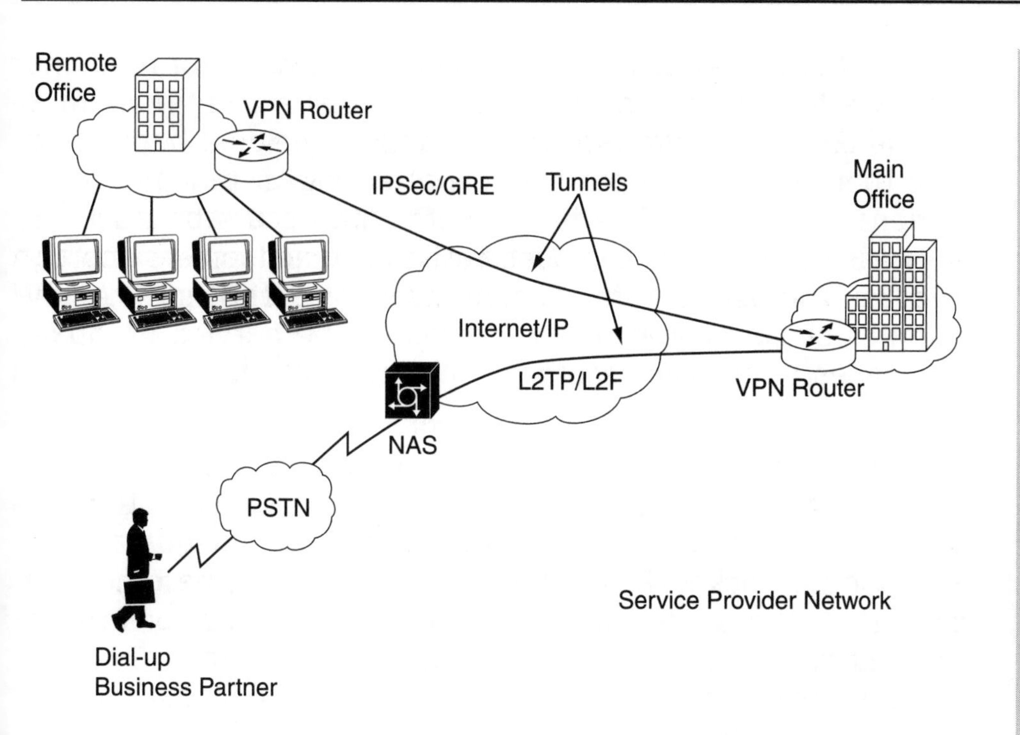

Figure 13.4. Dial-in access to a VPN

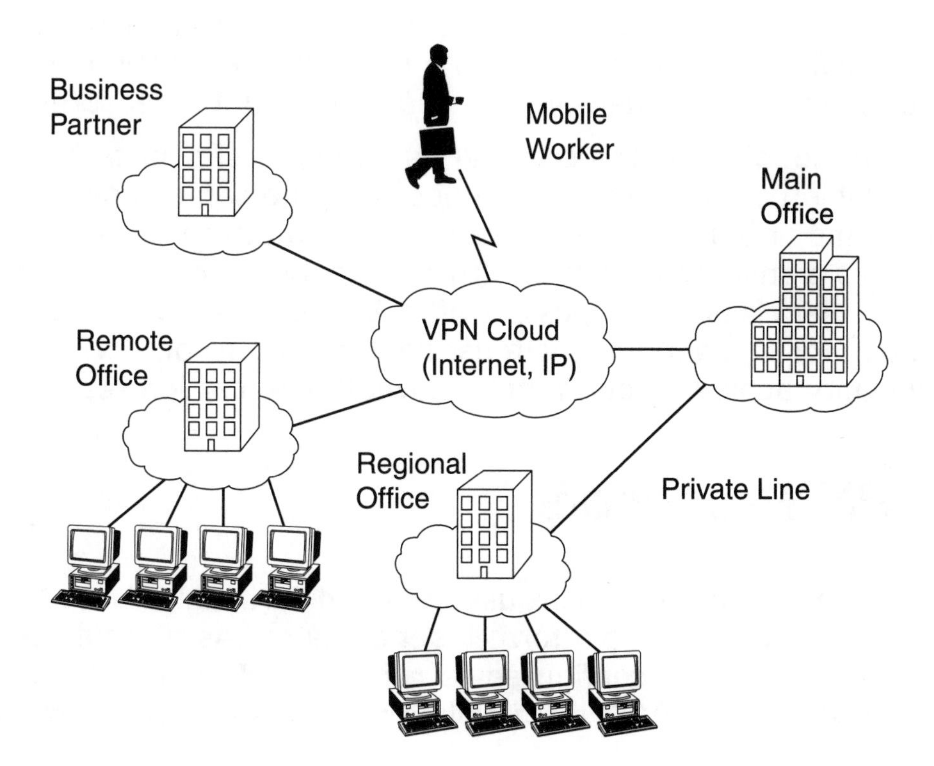

Figure 13.5. A hybrid private VPN

Cost savings occur through better data rates for modems. Because long-distance VPN users can dial a local modem at the office of the VPN carrier, the modem yields a better data rate than the throughput given with a long-distance or international direct call. The better date rate occurs because the VPN user can choose the best available local loop service and because the user can fully utilize available bandwidth. As the number of connected users increases, the service gradually decreases but never becomes completely blocked.

On the other hand, several important disadvantages remain for remote access VPN users. Unlike circuit-switched or leased-line data services, VPN tunnels over public routed networks may not offer any end-to-end throughput guarantees. Because the packets have a variable loss that may become high, the network may deliver the packets out of order or in fragmented condition. Because VPN connections attach first to a point of presence on the public network and then use the network to reach a remote peer before forming a private tunnel, the network may receive unsolicited data from other users of the public network. As a result, a remote access VPN may require more comprehensive and complex security measures.

Although remote access VPNs have the capability to use available bandwidth, VPNs do not have the same capabilities in terms of bandwidth reservation. Bandwidth reservation reserves transmission bandwidth on a network connection for particular classes or types of traffic. Because bandwidth reservation for in-bound traffic must have flow control, a VPN has difficulty achieving optimum bandwidth reservation results.

VPN Protocols

The Internet protocols used to build a Virtual Private Network also establish platform independence. As a result, any computer system configured to run on an IP network could become part of a VPN by installing remote software. Platform independence allows the implementation of Virtual Private Networks between corporations. VPN standards include the Point-to-Point Tunneling Protocol, Layer 2 Tunneling Protocol (L2TP), and IP Security Protocol (IPSec). The PPTP, L2TP, and IPSec specifications fit within the IPv6 Internet Protocol and support packet tunneling. In addition to the IPv6 specification, IPSec also operates within the IPv4 Internet Protocol.

PPTP and L2TP

The PPTP and L2TP protocols operate within multiprotocol environments but require additional support to deliver data privacy, integrity, and authentication. Unless supplemented with the IPSec, PPTP and L2TP cannot support Extranet applications. An Extranet requires keys and key management not seen with the PPTP and L2TP protocols.

VPNs and IP Tunnels

An IP tunnel encapsulates a data packet within a normal IP packet for forwarding over an IP-based network. The encapsulated packet may include an Internet Protocol or other protocols such as IPX, AppleTalk, SNA, or DECnet. Although the encapsulated packet does not require encryption and authentication, most IP-based VPNs, rely on encryption to ensure privacy and use authentication to ensure integrity of data. The use of encryption and authentication becomes especially important in VPNs accessible through the Internet.

Three basic types of IP tunnels exist. With IP tunnels between a remote user and a corporate firewall, the user's computer and the firewall control tunnel creation and deletion. When established between Internet Service Provider and a corporate, the ISP controls the tunnel creation and deletion. IP tunnels also exist between public Internet sites or over a service provider's IP network separated from the public Internet.

VPNs that function through the use of IP tunnels involve the self-deployment of the network. With this, users buy only the physical connections from an ISP, install VPN equipment, and establish any configuration or management. Along with self-deployed VPNs, ISPs, service providers, and other carriers also offer VPN services based on IP tunnels. Regardless of the service provider, these IP Tunnel VPNs offer fully managed services with options such as Service Level Agreements that ensure Quality of Service.

VPN/IP Tunnel Advantages and Disadvantages

VPNs based on IP tunnels offer several key advantages. As an example, replacing dedicated and long-distance connections with local connections reduces telecommunications costs. In addition, VPNs based on IP tunnels provide greater flexibility for the deployment of mobile computing solutions, telecommuting, and branch office networking. This flexibility becomes apparent when considering that easier eCommerce transactions and Extranet connections with business partners occur through the VPN/IP tunnel technologies. In addition, VPNs based on IP tunnels establish external Internet access and internal Intranet and Extranet access through a single secure connection.

Even though VPNs based on IP tunnels offer significant advantages, several disadvantages remain. VPN/IP tunnel solutions may have erratic QoS levels. Moreover, any VPN solution based on the public Internet requires higher levels of data security that include authentication and data encryption.

Virtual Private Network Classes

Class 0 VPN

A Class 0 VPN provides a simple, cost-effective solution for small companies with a single site and limited remote workers. Using the Point-to-Point Tunneling Protocol and packet filtering for Intranets, the implementation of a Class 0 VPN requires either Windows NT or a VPN-software solution installed on a server. Along with providing for e-mail and internal database access at a single site, Class 0 VPNs also allow file access for a maximum of 50 remote users through dial-up connections and Internet access at the main site through an DSL or fractional T1. Although Class 0 VPNs offer simplicity and low-cost implementation, the VPN solution lacks site-to-site flexibility and may require more maintenance than other VPN solutions.

Class 1 VPN

Class 1 VPNs offer the best solution for small to mid-sized companies that have multiple branch locations, access the Internet over T1 connections, and have a maximum of 250 remote users. In addition to those capabilities, Class 1 VPNs provide basic IPSec security with DES encryption and password authentication for remote access to e-mail and access to internal databases and files. Implementation of a Class 1 VPN occurs through a hardware VPN gateway with associated client software used for IPSec remote access. Like Class 0 VPNs, a Class 1 VPN combines easy design and installation with low cost. The use of IPSec security establishes greater security for both site-to-site and remote access while maintaining performance. When used with interoperable IPSec software, Class 1 VPNs support Extranet applications.

Class 2 VPN

Class 2 VPNs work best for medium-sized companies that may have a maximum of 500 remote users and 10 corporate sites. Because many medium-sized companies conduct high-value intellectual property transactions, Class 2 VPNs offer higher security through DES encryption, strong user authentication based on software tokens, and compatibility with firewalls. Scalability for Class 2 VPNs improves through the use of servers that manage the user names, passwords, and policies. Because Class 2 VPNs can offer network address translation, or NAT, the VPN can link privately addressed sites without requiring changes to the existing LAN addressing schemes. Although Class 2 VPNs provide higher security, site-to-site communication, and remote access, the VPN class has weaker support for Extranets and real-time applications.

Class 3 VPN

Medium to large companies rely on Class 3 VPNs to support thousands of remote users and hundreds of sites, to send and receive customer account status, and to complete supply chain transactions or eCommerce transactions. As an example, medical providers, insurance companies, manufacturers, or companies that operate as part of an extended supply chain may use Class 3 VPNs. Moving to figure 13.6, note that most Class 3 VPNs operate through a service provider that offers QoS and Service Level Agreements that support guaranteed response times for critical applications that run at a company site connected to the provider's backbone.

Class 3 VPNs use interoperable IPSec devices that enable Extranets for users who rely on different types of networks and equipment. User-level authentication occurs by employing authentication tokens or smart cards and supports secure remote access. To facilitate scalable administration, Class 3

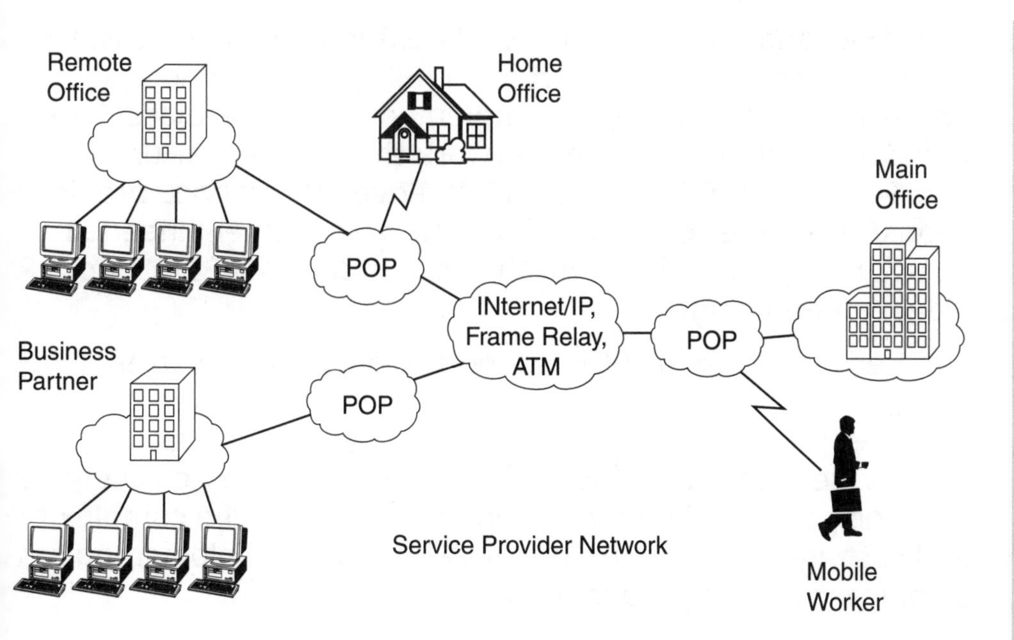

Figure 13.6. VPN access through a service provider

VPNs employ directory services for storing and retrieving policies as well as for storing the digital certificates used for authentication.

Each type of service requires an in-house certificate authority, outsourced certificate service, or some method of managing a public key infrastructure. Given the management complexity, the implementation of Class 3 VPNs requires higher level design skills, ongoing management, and larger amounts of time and resources. In addition, the implementation of Class 3 VPNs requires policies that address employee and partner access and overall system security.

Class 4 VPN

Class 4 VPNs present secure, flexible, and scalable networking solutions that can support secure transactions for large, multinational enterprises that have more than 10,000 users and thousands of sites. The VPNs extend chains of business partners and include support for a full range of remote access options. In most cases, the implementation of a Class 4 VPN relies on a service provider network that offers real-time QoS and Service Level Agreements. Moreover, the service provider should offer the level of bandwidth management needed to support applications such as voice over IP and IP videoconferencing.

Because Class 4 VPNs support sophisticated Extranet relationships that involve multiple — and sometimes international — partners, the Virtual Private Network class uses directory services and encryption technologies extensively to manage policy and verify user identities. Given the complexity and cost associated with Class 4 VPNs, the technology requires significant ongoing management.

Virtual Private Network Issues

Technological and practical issues can slow the implementation of a Virtual Private Network. For a VPN to function successfully, the network must provide essential features that solve problems resulting from the routing of private data across a shared public network. The features include security, performance, ease-of-management, scalability, efficiency, and cost-effectiveness. Because a VPN operates as a shared-access routed network, security becomes the main area of concern. As seen in chapter 12, good security practices require the use of encryption, the application of public and private keys, session and per-packet authentication, security negotiation, private address space confidentiality, and filtering.

Performance issues for a VPN occur because of the uniqueness of VPN traffic. Traffic from other than a tunnel end point can flood the receive path of a VPN from anywhere on the public network. If a VPN has provided the tunnel, the carrier can filter non-VPN traffic or provide a bandwidth reservation service. For client-based tunnels, however, those services do not exist.

VPN performance also depends on the speed of transmissions over either the Internet or a public IP backbone network and the efficiency of VPN processing at each end of the connection. VPN performance also depends on the speed of transmissions over either the Internet or a public IP backbone network and the efficiency of VPN processing at each end of the connection. IP datagrams sent across the VPN carrier service might experience packet loss and packet reordering. The PPP compression designed to increase the reliability of point-to-point links can cause packet loss to increase.

Encapsulation increases packet size because of the addition of information to each packet. In turn, routers may fragment the oversized packets and cause performance degradation. Packet fragmentation combined with data encryption can reduce dial-in system performance to unacceptable levels. Although data compression can alleviate the problem caused by encapsulation, a performance trade-off against the need for additional computing power will occur.

With bandwidth reservation, a VPN has the capability to reserve transmission bandwidth on a network connection for particular classes of traffic or particular users. When combined with QoS standards, bandwidth reservation also allows a VPN to allocate percentages of total connection bandwidth. With this, each traffic class or user may have a set of assigned priority levels, and a bandwidth reservation algorithm calculates the packets to drop when network traffic outgrows the available bandwidth.

Several options for bandwidth reservation in VPNs exist. On the in-bound side, an access device can apply tunnel and nontunnel bandwidth filtering techniques to a client's requirements. In addition, a VPN carrier can establish a predetermined capacity for each network link. Most VPN network servers or access concentrators can use in-bound, dynamic flow control for bandwidth reservation.

Applying encryption from host to host also changes the nature of IP-switching filters. For IP switching to function on encrypted data flows, the switch may use IPSec protocol headers to identify a communication stream. Switches that terminate secure tunnels cannot use fast forwarding because the encrypted IP packet must be reconstituted before forwarding.

VPNs and Business Opportunities

Companies that use a variety of data and voice services to meet their communication needs may find that VPNs and new technologies associated with VPNs will provide large-scale benefits. New customer-premises routers can operate as both Security Gateways and Multimedia Gateways while integrating a number of LAN and WAN capabilities such as hub and routing functions. The next generation routers also support applications such as Voice-over-IP (VoIP), IP-fax, and Internet access as well as VPN traffic over a single local-loop link to a service provider point of presence.

World Wide Web access and Web publishing may become the starting point for a company that wishes to take advantage of VPN services. Private remote access through VPN carrier networks may provide more cost-effective and scalable remote access to corporate networks. The use of a Virtual Private Network may also provide a common, bandwidth-efficient technology for adding new network links.

Virtual Local-Area Networks

A Virtual LAN can provide a solution for exponentially expanding connectivity for an organization. Even though a VLAN offers logical network segmentation based on workgroup func-

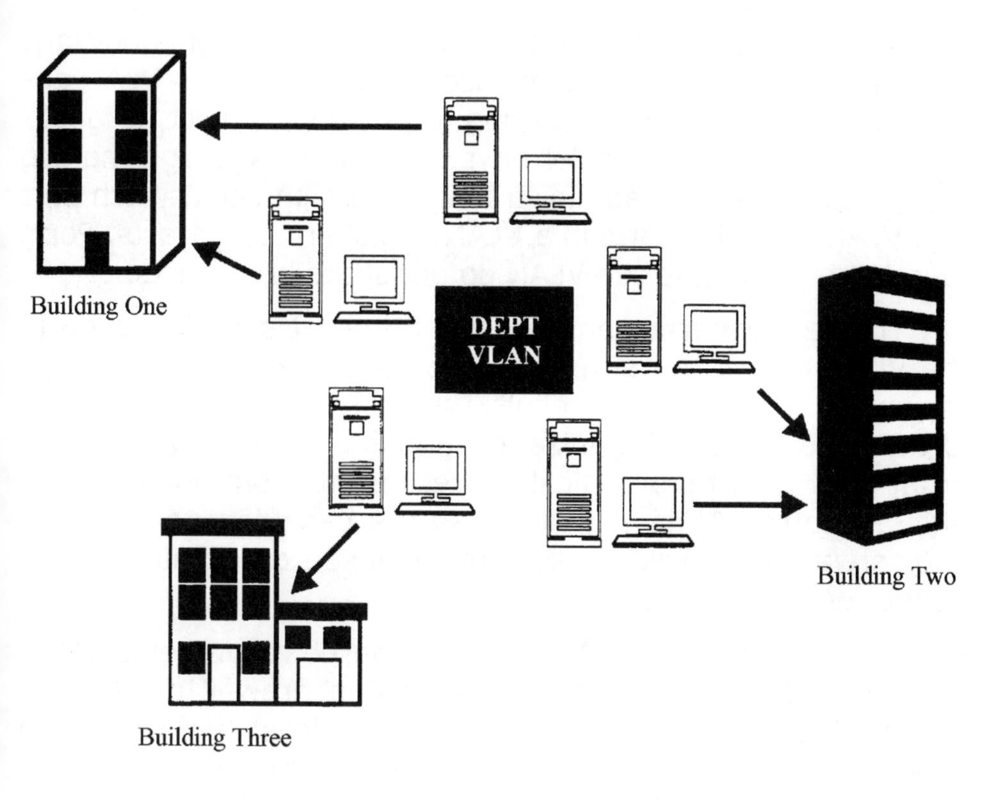

Figure 13.7. Diagram of a VLAN

tions rather than location and can enhance efficiency through the sharing of network resources, costs and configuration issues accompany the benefits. As figure 13.7 indicates, a VLAN creates the effect of a local-area network over a wide-area network by bundling access to distributed network resources.

In comparison to the traditional configuration of a local-area network, a VLAN gives users speed and security through switched access. A traditional LAN configures according to physical infrastructure and groups users according to locations. Routers provide segmentation for the network by interconnecting hubs. However, this type of segmentation does not accommodate groupings based on either workgroup association or a need for bandwidth. Users share the same segment and contend for the same bandwidth. Security issues occur because the segmentation through router ports prevents the logical assignment of network addresses across the network.

Switched VLAN configurations differ from LAN configurations because front-end hubs are replaced with switches. A VLAN enables the segmentation according to function, project, or application and does not consider physical locations through the use of a switch. A system administrator can assign each switch port to a VLAN, and the ports in a VLAN can share broadcasts. Ports that do not belong to the VLAN do not share broadcasts.

The Fast Ethernet and Gigabit Ethernet switches used in the VLAN act as the entry point for end-station devices into the switched fabric. As a result, the switch facilitates communication across the organization and provides intelligence to group users, ports, or logical addresses into common communities of interest. Each switch has the capability for filtering and forwarding decisions by packet based on administrator-defined VLAN metrics.

Routers provide VLAN access to shared resources such as servers and hosts. Along with providing firewall protection, routers used within a VLAN establish broadcast suppression to policy-based control, broadcast management, and route processing and distribution. In addition, routers connect to other parts of the network that remain logically segmented according to location and subnet.

VLAN Standards

The IEEE has recently authored two standards for VLAN management. The 802.1p standard improves support of time-critical and multicast-intensive applications such as applications that involve the use of interactive multimedia. In addition, the 802.1p standard addresses VLAN set-up issues and VLAN traffic. The 802.1q standard establishes an architecture and provides protocols and mappings for bridges to provide interoperability and consistent management. Moreover, the 802.1q standard describes a standard method for using a VLAN within a frame-based network.

VLAN Configurations

As the uses for VLANs have grown, several different configurations have become popular. With the Authenticated User VLAN, a server authenticates users before granting access to network resources. A Layer 3 VLAN allows the assignment of traffic with specific protocol criteria, and a MAC Address VLAN groups MAC addresses into a broadcast domain. The MAC Address VLAN offers security and control through MAC address identification.

Organizations use a Multicast VLAN configuration for videoconferencing and news feeds to workstations. When users open applications that contain a multicast, they dynamically join the multicast VLAN associated with the application. Closing the application disconnects the user from the multicast.

Two VLAN configurations occur around policies, but a third has a basis on the assignment of ports. A Policy-based VLAN allows network or system administrators to use any combination of VLAN policies to create a VLAN that meets the requirements of the administrator. As a result, a Policy-based VLAN can implement different assignment methods including MAC source addresses and Layer 3 addresses. A Protocol Policy VLAN configuration uses the protocol criteria within a frame and allows an administrator to specify the criteria for creating the VLAN and to group devices into the VLAN based on the policy. The Port-based VLAN configuration assigns devices to the VLAN based on the physical attachment to a port.

Implementing a VLAN

The implementation of a VLAN can reduce obstacles to departmental networking that occur because of physical boundaries between resources. A VLAN eliminates the need for physical subnets by consolidating the resources. When moving to implement a VLAN in an organization, always consider the need for access, the assignment of IP addresses, the configuration of links, the configuration of trunks, the need for Inter-VLAN communication, security issues, and traffic requirements. Because new networking equipment is being used, increased costs can accompany the VLAN solution in the form of costs per megabit of data transfer through the switch.

Access to the VLAN should occur according to criteria based on an organization's needs such as department groupings. The assignment of IP addresses becomes simplified through the assignment of network addresses with the same subnet to all stations residing on the same VLAN. Although the configuration of links must occur during the assignment of ports, trunk configurations must accommodate more than one VLAN.

Communication between VLANs may occur because a VLAN user needs to access resources from another VLAN. As a result, the router will require a configuration that allows translational bridging. The configuration between VLANs must also consider media type as well as the number of users and the amount of traffic. Because a VLAN groups users, a system administrator should consider user groups that restrict access to proprietary data. In addition, a group working with sensitive data should remain on a single VLAN.

14 Network Security

Introduction

A banking executive once said, "The information we have on money is more valuable than the money itself." Adding to this statement, modern computing systems contribute to the control/conduct of business and services at all levels of an organization. As a result, data have become more valuable than ever. From many perspectives, the focus of information technology planning has changed from the computer and its software to the processing, storage, and retrieval of data.

Data corruption or loss can occur because of many factors including hardware and software failures, natural disasters such as floods or earthquakes, fire, theft, vandalism, human error, intentional erasure/corruption, and employee malice. This chapter introduces you to threats to system security and defines policy measures for blocking those threats. Security issues listed within the module include virus protection, the use of electronic mail, the need for a routine system backup, and the need for firewalls.

The Need for System Security

No one thinks seriously about computing and ignores communication. The world of computing has become a world of connectivity and the sharing of data and group work. With this, the dangers of insecure communications are relatively obvious. As we make greater use of different technologies as a part of the vital intra- and interorganizational communications system at any business, we must also rely on the integrity, privacy, and reliability of the communications taking place.

Unfortunately, a fundamental flaw exists in the way that most of us think in terms of security. Generally, when we consider computer security, we think only of issues such as access control, user rights, and privileges. Along with considering those issues, we should also analyze what we need to secure and create a model that helps to classify what to protect.

Building this model involves teamwork between systems administrators and all professionals within the organization in an effort to identify objectives. However, rather than thinking in terms of the number of calls answered during a certain time period, high-level objectives should be reduced to tasks by asking, "What are the processes they go through to achieve the objectives." Given this information and with their continuing help, you can identify the data and the information systems needed to deliver the tasks. This type of approach to system security will make it easier for everyone to accept and support security measures.

Information Technology Crimes

An information-technology crime is an illegal act perpetrated against computers or telecommunication systems. In addition, information-technology crimes involve the use of computers or telecommunications to accomplish an illegal act.

Roughly speaking, security issues deal with personal information that is not voluntarily given, such as items stored on one's computer hard drive. Furthermore, security covers information contained in e-mail messages or other electronic transactions that are intercepted between their source and destination. The interception of credit card numbers by hackers during on-line business transactions is an example of a security breach.

Hackers

Hackers are amateur computer enthusiasts or programmers who lack formal training. Depending on how it is used, the term can be either complimentary or derogatory. However, the term is developing an increasingly derogatory meaning. In this sense, the term refers to individuals who gain unauthorized access to computer systems for the purpose of stealing and corrupting data.

Viruses

A virus is a man-made piece of code that can replicate itself. As you may know, several thousand types of viruses exist. The viruses may be passed from a floppy disk to a hard disk on a microcomputer system or may be downloaded from the Internet. In some cases, viruses will cause simple errors such as formatting problems. In other cases, a virus may erase the content of a hard disk drive.

A simple virus that can make a copy of itself is relatively easy to produce. Even such a simple virus is dangerous because it will quickly use all available memory and bring the system to a halt. An even more dangerous type of virus can transmit itself across networks and bypass security systems. A special type of virus called a worm can replicate itself and use memory, but it cannot attach itself to other programs.

Viruses can spread through software installations, as macros embedded in e-mail messages, through the use of unauthorized software, and through files downloaded from the World Wide Web. The following list provides a sampling of how viruses attacked corporate networks.

- A manufacturer of CD-ROMs shipped a copy of a dangerous polymorphic virus worldwide on CD. Users around the world were infected.
- A large insurance organization rented 15 computers from a local firm; all the computers arrived infected with a virus named SatanBug. The virus quickly spread to their servers and other workstations in the organization.
- A network consulting firm arrived at the door of a small company to help it with its network problem. When the first disk was placed in a drive, anti-virus software warned that the disk was infected. It was discovered that all 15 disks the consultants arrived with were infected, but with three different boot viruses. The consultants claimed to have checked their disks with an anti-virus product.
- The training department in one of the world's largest organizations infected every student who attended courses with a virus named LBB Stealth.

Electronic Mail

E-mail has become one of the premier forms of electronic communication in business today. Corporations use electronic mail to take orders, make purchases, pass along instructions, confer about delicate transactions, exchange employee and customer information, and do, in short, just about everything usually accomplished with the telephone, interoffice mail, and face-to-face meetings at the water cooler.

However, e-mail makes the water cooler look like an encrypted CIA data link. Although we use e-mail applications to move data and communicate freely with others, we should remain aware that anyone can look at an e-mail file. In addition, we should also remember that an e-mail message remains on mail servers and on hard disks at both the sender and receiver sites.

Consistent e-mail policies inform everyone in an organization about what they can and cannot do with electronic mail. E-mail exists as only one type of electronic document that flows in and out of the organization. In addition to e-mail messages, uploaded and downloaded text, graphics, and images can flow in and out of computer systems. Each offers the potential for serious problems. As an example, if a technician reads pornography or racist material from the World Wide Web on a system belonging to the organization, the potential for liability exists. Moving to another example, if an employee uses a corporate mail server to send harassing or threatening e-mail, then the corporation will remain liable for those actions.

In the News...

Cyber Attacks Rise from Outside and Inside Corporations

Dramatic Increase in Reports to Law Enforcement

SAN FRANCISCO — The Computer Security Institute (CSI) announced today the results of its fourth annual "Computer Crime and Security Survey." The "Computer Crime and Security Survey" is conducted by CSI with the participation of the San Francisco Federal Bureau of Investigation (FBI) Computer Intrusion Squad. The aim of this effort is to help raise the level of security awareness as well as determine the scope of computer crime in the United States. Highlights of the "1999 Computer Crime and Security Survey" include the following:

- *Corporations, financial institutions, and government agencies face threats from outside as well as inside.*
- *System penetration by outsiders increased for the third year in a row; 30 percent of respondents report intrusions.*
- *Those reporting their Internet connection as a frequent point of attack rose for the third straight year; from 37 percent of respondents in 1996 to 57 percent in 1999.*

Meanwhile, unauthorized access by insiders also rose for the third straight year; 55 percent of respondents reported incidents. Other types of cyber attack also rose. For example, 26 percent of respondents reported theft of proprietary information. Perhaps the most striking result of the 1999 CSI/FBI survey is the dramatic increase in the number of respondents reporting serious incidents to law enforcement: 32 percent of respondents did so, a significant increase over the three prior years, in which only 17 percent had reported such events to the authorities.

For the third straight year, financial losses due to computer security breaches mounted to over a $100,000,000. Although 51 percent of respondents acknowledge suffering financial losses from such security breaches, only 31 percent were able to quantify their losses. The total financial losses for the 163 organizations that could put a dollar figure on them added up to $123,779,000.

The most serious financial losses occurred through theft of proprietary information (23 respondents reported a total of $42,496,000) and financial fraud (27 respondents reported a total of $39,706,000).

Although these survey results indicate a wide range of computer security breaches, perhaps the most disturbing trend is the continued increase in attacks from outside the organization. This trend was reinforced by other survey results. For example, of those who acknowledged unauthorized use, 43 percent reported from one to five incidents originating outside the organization, and 37 percent reported from one to five incidents originating inside the organization.

Further evidence of increased system penetration from the outside can be gleaned from a series of questions on WWW sites and electronic commerce that were asked for the first time this year. Ninety-six percent of respondents have WWW sites, 30 percent provide electronic commerce services. Twenty percent had detected unauthorized access or misuse of their WWW sites within the last 12 months (disturbingly, 33 percent answered "don't know.") Of those who reported unauthorized access or misuse, 38 percent reported from two to five incidents, and 26 percent reported 10 or more incidents. Thirty-eight percent reported that the unauthorized access or misuse came from outside. Several types of attack were specified:

- *98 percent reported vandalism,*
- *93 percent reported denial of service,*
- *27 percent reported financial fraud,*
- *25 percent reported theft of transaction information.*

Only 12 of the 95 respondents who had their WWW sites attacked could quantify their financial losses. The total losses for the 12 respondents totaled $2.383 million (an average of $198,583 in financial losses for each respondent.) Sixty-two percent of respondents reported computer security breaches within the last twelve months.

The breaches detected by respondents include a diverse array of serious attacks, several of which rose in the number of reports from 1998 to 1999;

for example, system penetration by outsiders, unauthorized access by insiders, and theft of proprietary information as mentioned earlier.

Here are some other examples.

- *Denial of service attacks was reported by 32 percent.*
- *Sabotage of data or networks was reported by 19 percent.*
- *Financial fraud was reported by 14 percent.*

Insider abuse of Internet access privileges (for example, downloading pornography or pirated software or engaging in inappropriate use of e-mail systems) was reported by 97 percent.

- *Virus contamination was reported by 90 percent.*
- *Laptop theft was reported by 69 percent.*

Managing Security

Rather than remain reactive about security issues, organizations must take a proactive approach to making policies about secure technology practices. In some cases, achieving a policy goal requires a change in the way that we view knowledge, experience, personal values, and beliefs. In addition, successful security policies should link with business policies. With no formal policy goals for security, informal practices and procedures will dominate the organization. With this, we should carefully document the policies and seek formal management approval and support.

Policy is a framework to manage change in the context of business strategy. Moreover, a good policy works as an instrument for managing risk. Good security policies depend on the integration of three key components:

- organization,
- practice, and
- technology.

Many of us wonder how security professionals can easily and quickly pinpoint strengths and weaknesses in security. In brief, professionals concerned with system security consider how a security threat will affect each of the components. Mixing the components makes it possible to create security solutions that meet organizational needs.

More than likely, we cannot attain 100 percent system security. For example, attacks by the Defense Information Systems Agency, or DISA, on 9,000 Department of Defense computer systems had an 88 percent success rate but were detected by less than 1 in 20 of the target organizations. Of those organizations, only 5 percent actually reacted to the attack. However, the use of virus protection software, firewalls, and other security measures will block most threats.

Designing an effective security plan for your organization requires an understanding of how data travels through the network. The most secure transactions occur from end to end and only between the parties involved in the transaction. Good network security hinges on confidentiality, or knowing that the data has remained private; integrity, or knowing that someone has not manipulated the data; authentication, or remaining sure of the sender and receiver; and nonrepudiation, or knowing that the transaction stands without denial. Confidentiality, integrity, authentication, and nonrepudiation match with the technologies associated with encryption, digital signatures, and certificates.

Planning for a Secure System

Security planning in a technology environment identifies the technology assets that need protection; quantifies values; and determines threats. For example, we know that corporate files about client relations have value and need protection. From there, we can determine that the lack of password or virus protection for that information allows a threat to exist.

Rules-Based Management

We can take a quantitative approach to the security of our technology systems. However, a quantitative approach only benefits organizations that have the time and available resources to complete such a project. During a quantitative analysis, a technology team will categorize and consider every possible asset and threat to the system. Quantitative analysis requires the collection of statistical information about each possible type of threat involved for each asset. For example, with a quantitative analysis, one end result contained in the report may resemble, "There are 242 incorrect password entries logged on system 'green' each month." Before applying the quantitative approach, though, consider that a technology team will require at least three months for an analysis of staff within the organization.

We can also take a qualitative approach to system security. Qualitative methods provide a snapshot to examine the critical business assets of an organization. Performing a qualitative analysis requires a subjective assignment of threat categories (high, medium, low) to threats affecting critical business systems. Organizations that rely on quick, fluid decision making can benefit from a purely qualitative analysis. Additionally, because qualitative analyses require fewer people and much less time to finish, organizations can employ a qualitative analysis several times a year.

To make the most of our efforts, we can blend quantitative and qualitative aspects together to fit the organizational requirements of the business. This type of analysis will collect detailed threat statistics about critical business assets while still evaluating other less critical risks with a qualitative approach. The amount of blending of quantitative and qualitative information should align with organziational goals.

Authentication

Authentication tools and policies provide the system-wide capability to recognize and verify the identity of users. As already mentioned, one portion of authentication occurs through the use of user names and passwords. Many systems combine the use of passwords with other techniques such as the use of smart cards, biometrics, voice recognition, or digital signatures.

A password is a secret series of characters that allows an individual to access a file, computer, or program. In addition, data files and programs may require a password. On multiuser systems, each user must enter his or her password before the computer will respond to commands. The password helps ensure that unauthorized users do not access the computer.

Most security schemes provide several increasingly restrictive levels of access. For example, most networks request a password from users as they attempt to log on. Even though the software on the server may have one set of passwords for all users, it will always feature a higher level password for the administrator. This attribute is necessary because of the amount of control given to the administrator. A password should be something that is difficult to guess. In practice, however, most people choose a password that is easy to remember — such as their name or their initials. For that reason, it may be relatively easy to break into some computer systems.

When establishing your password for either a stand-alone system or for network access, follow these simple rules:

- Make your password at least eight characters long. The longer password is much more difficult to guess.
- Never use an obvious set of characters such as your name, birth date, or the name of the program. Instead, choose an obscure name or phrase. Use at least one space or nonalphabetic character if the program allows the option.

- Never tell anyone your password.
- Never write the password down.
- Change your password often.
- Never embed your password into a script or macro file.

Two basic approaches — called PAP and CHAP — to user name/password authentication exist. The Password Authentication Protocol, or PAP, asks the requester to provide a password. If the system includes the password in a set of user profiles, the requester gains access to the system. The Challenge Handshake Authentication Protocol, or CHAP, challenges the individual requesting access to encrypt a response to the randomly generated challenge string that accompanies the server hostname. In turn, the client uses the hostname to look up an appropriate key, combines the key with the challenge, and encrypts the combination. Then, the resulting string and client hostname return to the server.

After the server performs the same computation on the challenge string presented by the client, it allows the client to connect only if it receives an indentical result. As a result, the system enforces a different password for each entry. Many systems combine the CHAP system with a physical key — such as a smart card — for the encoding of the challenge message.

Nonrepudiation

Nonrepudiation protects against the sender or receiver denying that they sent or received certain communications. The three types of nonrepudiation services for computer messaging include:

- Nonrepudiation of Delivery Service,
- Nonrepudiation of Origin Service, and
- Nonrepudiation of Submission Service.

Nonrepudiation of Delivery Service provides the sender with proof of successful message delivery to the intended recipient. As an example, an electronic mail package often includes the option of requesting a return receipt. Nonrepudiation of Origin of Service provides the recipient with proof the originator of the message and the type of message content. Nonrepudiation of Submission Service offers proof that about the sending of a particular message from a specific sender.

Access Control

Access control methods either provide or reject access to some service or data in a system. The service may consist of software, hardware, an operating system, a file system, data, a collection of files, or a combination of items. An organization defines a set of access rights, privileges, and authorizations and assigns those to appropriate individuals working under the organization.

Levels of access control include public access, private access, and shared access. Public access allows all users to have read-only access to file information. Private access provides read- and write-access to specific users, and shared access allows all users to read from and write to files. Restrictions on access may include passwords, encryption, firewalls, callback systems, and identification systems.

Public-Key Infrastructure

Encryption translates data into a secret code and provides the most effective method for achieving data security. To read an encrypted file, an individual must have access to a secret key or password that enables the deciphering and reading of the file. We can describe unencrypted data as plain text and

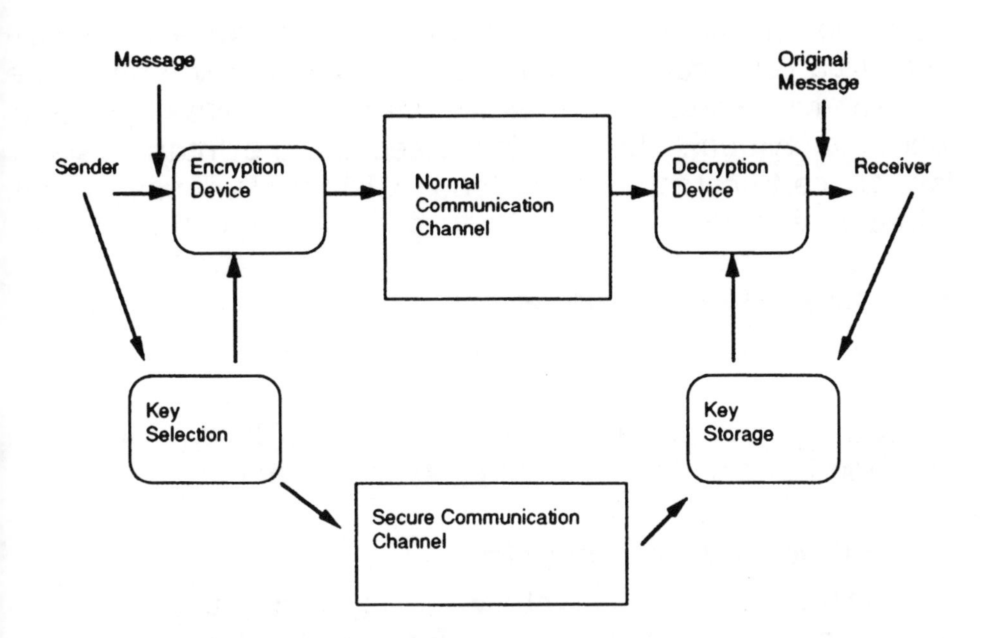

Figure 14.1. Flow chart of encryption and decryption

encrypted data as cipher text. The security of encrypted data depends on the number of bits within the algorithm and key. A larger number of bits translates into more computing power required to break the key. Figure 14.1 shows a flow chart for an encryption and decryption sequence.

Two main types of encryption are asymmetric, or public-key encryption, and symmetric encryption, or private-key encryption. Public-key encryption applies a publicly known key to encrypt data and a second private key to decrypt the data. Private-key encryption methods use a single-key algorithm known only by the sender and receiver. Although each method remains effective only as long as the key stays secure, public-key encryption offers an advantage in that it never requires the transmission of the private key. Private-key encryption systems require the transmission of the secret key to the recipient before decryption can begin.

Public-Key Infrastructure, or PKI, describes the method chosen for generating and distributing keys. Because of the difficulty seen with generating and distributing keys, PKI binds a key to a certificate. Rather than relying directly on encryption, the certificate confirms that the public key belongs to a specific individual. The X.509 has become the Internet standard for certificates but may require 10 kb of memory when stored on a card.

Data Encryption Standard

Developed by IBM, the Data Encryption Standard, or DES, allows a single key to encrypt and decrypt data. DES has received the most testing of any encryption/decryption standard and remains as a federal encryption standard.

RSA Public-Key Cryptosystem

The RSA Public-Key Cryptosystem exchanges authenticated secret messages without exchanging secrets. Rather than using the same key for the encryption and decryption of data, the RSA system uses a matched pair of encryption and decryption keys. Each key transforms the data in one direction.

The RSA system relies on the Public-Key Cryptography Standards, or PKCS, and that provides compatibility between vendors. In turn, the PKCS covers a set of specifications that describe methods for implementing public-key security. Any platform that conforms to PKCS standards can receive, authenticate, and decrypt messages created and sent from another application that complies with the standards.

Kerberos

Kerberos maintains the secrecy of every user's key by storing copies on a secure server and authenticating the identity of every user and network service. If one key or server becomes compromised, all keys change. Kerberos has become an Internet standard and a standard for remote authentication in client-server environments.

Security Protocols

Organizations such as the Internet Engineering Task Force, or IETF, have also begun to apply security measures at the IP layer of the Internet. Because this protection extends to the lowest possible protocol layers of the architecture, it can protect against upstream as well as downstream applications and prevent connections between unauthorized systems. Working within the IETF, the Internet Protocol Security group, or IPSEC, produced the Simple Key Management for Internet Protocol, or SKIP, as a technique for providing authentication and encryption security at the Internet IP layer.

Simple Key Management for Internet Protocol

SKIP relies on the existence of an authority residing within the network that can issue a certificate to known trusted entities within the system. If an entity claiming membership in the system requests an action, the receiving computer system can have the requester present an encrypted certification as verification. The certificate usually consists of a secret encoding technique and a secret key only available to authorized users.

Secure Socket Layer Protocol

Developed by Netscape, the Secure Socket Layer, or SSL, Protocol runs at the Transport Layer of the OSI model. With application programs running on top of it, SSL relies on public-key cryptography for encryption and Digital ID certificates for authentication. With the cryptography enabled, users can transmit secure information safely. The Digital ID certification provide confirmation about the identity of the sender. The Secure Socket Layer Protocol will run only if the Web server software runs on a port other than the standard TCP/IP port. Uniform resource locators based on the SSL Protocol begin with https:// rather than http://.

Secure Hypertext Transfer Protocol

The Secure HyperText Transfer Protocol, or S-HTTP, operates at the server and provides the capability for secure Web transactions. S-HTTP extends the HyperText Transfer Protocol for authentication and data encryption between a Web server and a Web browser by encrypting Web traffic between a client and server page by page. By adding cryptography at the application layer, S-HTTP ensures the security of end-to-end transactions. Uniform resource locators for Web sites utilizing S-HTTP begin with shttp:// rather than http://.

Firewalls

As shown in figure 14.2, a firewall consists of a system or group of systems that enforces an access control policy between two networks. Effective firewalls can present just a single IP address to the outside world and hide the real structure of a network from prying eyes through a process called address translation. Unregistered addresses map to a single, legal address. In addition, a firewall also provides full auditing and reporting facilities. A firewall can protect a corporate network from unauthorized external access via the Internet, but it can also prevent unauthorized internal access to a workgroup or LAN within a corporate network.

Most firewalls isolate an internal network by either denying the use of unsafe services on an Internet server or use packet filtering to prevent traffic from passing to the internal network from anywhere but authorized sites. A packet filter

Figure 14.2. A firewall appliance

checks IP address and service information to determine the source of incoming data traffic. Because of the transparency associated with packet filtering, users see no change in network performance and have no password requirements. Packet filtering provides a single point of entry and exit and disables services accidentally initiated on network equipment.

Securing Remote Access Connections

Callback Security Systems

Callback security systems control the remote dial-up access to a network through modems. During operation, the security system uses an interactive process between a sending and receiving modem. First, the answering modem requests the caller's identification. Then, the modem disconnects the call and verifies the caller's identification against a user direct. After the system verifies the identification, the answering modem calls back the newly authorized modem at a number that matches the caller's identification. As a result, communication occurs only between authorized devices.

Most network administrators combine callback security systems with password identification security. With this additional security method, the system verifies a password, disconnects the in-bound call, and calls the remote user at a predetermined telephone number. At this point, both modems enter the "pass data" mode, and communication begins. The combination of security methods always results in a callback to a secure location.

Remote Authentication Dial-In User Service

Remote Authentication Dial-in User Service, or RADIUS, has become one of the more popular public network authentication protocols. As the name shows, RADIUS has a primary purpose of offering centralized access control for remote dial-in users. The security solution simplifies security administration by placing passwords, user names, profiles for remote users, and other security and accounting information in a central server. When a request for network access occurs, RADIUS challenges the user.

Preparing for Viruses

Since 1987, when a virus infected ARPANET — a large network used by the Defense Department and many universities — many anti-virus programs have become available. An anti-virus program periodically searches a hard disk for viruses and removes any that are found. Many different types of excellent virus-checking software packages exist on the market. Those packages automatically check inserted floppy disks, the hard disk drive, and the computer memory and scan any files downloaded from the Internet. To protect the organization from a virus, never install untested software on your computer.

Virus Protection Policies

When planning for different technologies, virus protection policies frequently need to be created. The policies that are created may resemble the activities in the following list:

- Select a core group of virus response team members from across the organization.
- Establish that all users should run a smart anti-virus behavior blocking device driver at all times.
- Periodically ensure that you're the selected anti-virus device driver is still running as it was originally installed.

- Ensure that the latest version of the virus-scanning software is running on the computer system.
- Install products that offer intelligent signaling.
- Install an anti-virus software package on servers and perform real-time scans.
- Scan files as they are opened, copied, or renamed.
- Do not allow the use of pirated software in the organization.
- Develop a list of authorized software, including version information.
- Ask users to verify that all disks and files are virus-free.

Index

Symbols

A

B

C

D

E

F

G

H

I

J

K

L

M

N

O

P

Q

T

U

V

W

X

EXPLORING LANS FOR THE SMALL BUSINESS & HOME OFFICE

Author: LOUIS COLUMBUS
ISBN: 0790612291 • **SAMS#:** 61229
Pages: 304 • **Category:** Computer Technology
Case qty: TBD • **Binding:** Paperback
Price: $39.95 US/$63.95CAN

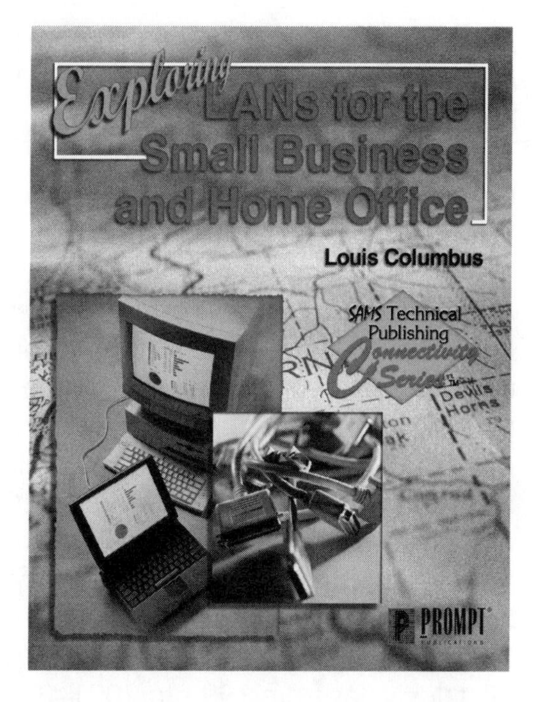

About the book: Part of Sams Connectivity Series, *Exploring LANs for the Small Business and Home Office* covers everything from the fundamentals of small business and home-based LANs to choosing appropriate cabling systems. Columbus puts his knowledge of computer systems to work, helping entrepreneurs set up a system to fit their needs.

PROMPT® Pointers: Includes small business and home-office Local Area Network examples. Covers cabling issues. Discusses options for specific situations. Includes TCP/IP (Transmission Control Protocol/Internet Protocol) coverage. Coverage of protocols and layering.

Related Titles: *Administrator's Guide to E-Commerce,* by Louis Columbus, ISBN 0790611872. *Administrator's Guide to Servers,* by Louis Columbus, ISBN

Author Information: Louis Columbus has over 15 years of experience working for computer-related companies. He has published 10 books related to computers and has published numerous articles in magazines such as *Desktop Engineering, Selling NT Solutions,* and *Windows NT Solutions.* Louis resides in Orange, CA.

RCA/GE/PROSCAN TV MISCELLANEOUS SERVICE ADJUSTMENTS

Author: SAMS TECHNICAL PUBLISHING
ISBN: 0790612429 • **SAMS#:** 61242
Pages: 336 • **Category:** Troubleshooting & Repair
Case qty: TBD • **Binding:** Paperback
Price: $34.95 US/$55.95CAN

About the book: Sams Technical Publishing's Engineering Staff has scoured their databases and come up with the perfect reference book for the shop and a MUST for the travling service technician! *RCA/GE TV Miscellaneous Service Adjustments* is a compilation of Miscellaneous Service Adjustments including Factory On Screen Menu settings on the newer sets found in PHOTOFACTS covering RCA/GE televisions from 1994 to 2001. Covering over 530 models, this gathering of facts, figures, adjustments and other information will be a tool that every service technician wants to have in his or her toolbox!

Prompt Pointers: Allows a service technician to carry important information grouped by manufacturer. An excellent tool for technicians of any level. An essential tool for in-home repairs.

Related Titles: *Zenith TV Miscellaneous Service Adjustments*, ISBN 0790612445. *Manufacturer to Manufacturer Part Number Cross Reference with CD-ROM*, ISBN 0790612321. *Semiconductor Cross Reference Guide 5E*, ISBN 0790611392.

Author Information: Sams Technical Publishing is the leader in the publishing of service documentation and schematics for TV repair. Since 1946 Sams Technical Publishing has produced PHOTOFACT® service documentation for the TV repair technician.

APPLIED SECURITY DEVICES & CIRCUITS

Author: PAUL BENTON
ISBN: 079061247X • **SAMS#:** 61247
Pages: 280 • **Category:** Projects
Case qty: TBD • **Binding:** Paperback
Price: $34.95 US/$55.95CAN

About the book: The safety and security of ourselves, our loved-ones and our property are uppermost in our minds in today's changing society. As security components have become user-friendly and affordable, more and more people are installing their own security systems. Paul Benton covers this topic in a "secure" way, applying proven electronics techniques to do-it-yourself security devices.

Prompt Pointers: Includes automobile security systems, basic alarm principles, and high-voltage protection. Outlines over 100 applied security applications. Contains over 200 illustrations.

Related Titles: *Guide to Electronic Surveillance Devices*, ISBN 0790612453, *Guide to Webcams*, ISBN 0790612208, *Applied Robotics*, ISBN 0790611848.

Author Information: Paul Benton has been involved in electronics since leaving school originally as a TV and radio technician, before becoming involved in electronic security devices and techniques in the 1980s'. Under the name of Paul Brookes, his mothers' maiden name, Benton has written a number of electronics related books and articles. As a teacher and lecturer at the university level, Benton remains current with today's technologies and currently works for an international electronic company in England.

ADMINISTRATOR'S GUIDE TO SERVERS

Author: LOUIS COLUMBUS
ISBN: 0790612305 • **SAMS#:** 61230
Pages: 304 • **Category:** Computer Technology
Case qty: TBD • **Binding:** Paperback
Price: $39.95 US/$63.95CAN

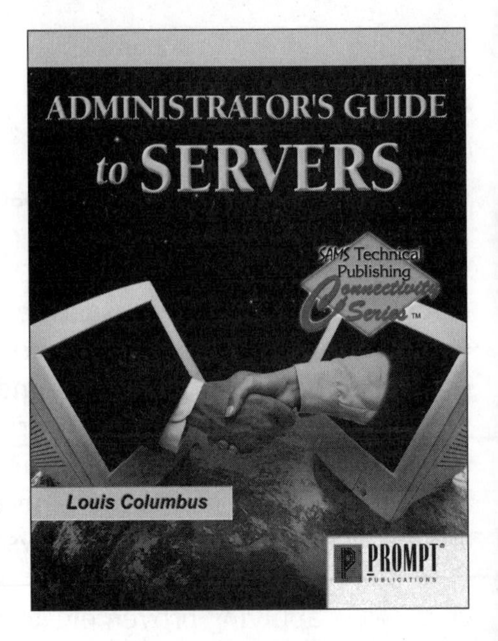

About the book: Part of Sams Connectivity Series, *Administrator's Guide to Servers* piggy-backs on the success of Columbus' best-selling title *Administrator's Guide to E-Commerce.* Columbus takes a global approach to servers while providing the detail needed to utilize the correct application for your Internet setting.

PROMPT® Pointers: Compares approaches to server development. Discusses administration and management. Balance of hands-on guidance and technical information.

Related Titles: *Administrator's Guide to E-Commerce,* by Louis Columbus, ISBN 0790611872. *Exploring LANs for the Small Business and Home Office,* by Louis Columbus, ISBN 0790612291. *Computer Networking for the Small Business and Home Office,* by John Ross, ISBN 0790612216.

Author Information: Louis Columbus has over 15 years of experience working for computer related companies. He has published 10 books related to computers and has published numerous articles in magazines such as *Desktop Engineering, Selling NT Solutions,* and *Windows NT Solutions.* Louis resides in Orange, CA.

ZENITH TV MISCELLANEOUS SERVICE ADJUSTMENTS

Author: SAMS TECHNICAL PUBLISHING
ISBN: 0790612445 • **SAMS#:** 61244
Pages: 300 • **Category:** Troubleshooting & Repair
Case qty: TBD • **Binding:** Paperback
Price: $34.95 US/$55.95CAN

About the book: Sams Technical Publishing's Engineering Staff has scoured their databases and come up with the perfect reference book for the shop and a MUST for the travling service technician! *Zenith TV Miscellaneous Service Adjustments* is a compilation of Miscellaneous Service Adjustments including Factory On Screen Menu settings on the newer sets found in PHOTOFACTS covering Zenith televisions from 1994 to 2001. Covering over 295 models, this gathering of facts, figures, adjustments and other information will be a tool that every service technician wants to have in his or her toolbox!

Prompt Pointers: Allows a service technician to carry important information grouped by manufacturer. An excellent tool for technicians of any level. An essential tool for in-home repairs.

Related Titles: *RCA/GE TV Miscellaneous Service Adjustments*, ISBN 0790612429. *Manufacturer to Manufacturer Part Number Cross Reference with CD-ROM*, ISBN 0790612321. *Semiconductor Cross Reference Guide 5E*, ISBN 0790611392.

Author Information: Sams Technical Publishing is the leader in the publishing of service documentation and schematics for TV repair. Since 1946 Sams Technical Publishing has produced PHOTOFACT® service documentation for the TV repair technician.

ADMINISTRATOR'S GUIDE TO DATAWAREHOUSING

Author: AMITESH SINHA
ISBN: 0790612496 • **SAMS#:** 61249
Pages: 304 • **Category:** Computer Technology
Case qty: TBD • **Binding:** Paperback
Price: $39.95 US/$63.95CAN

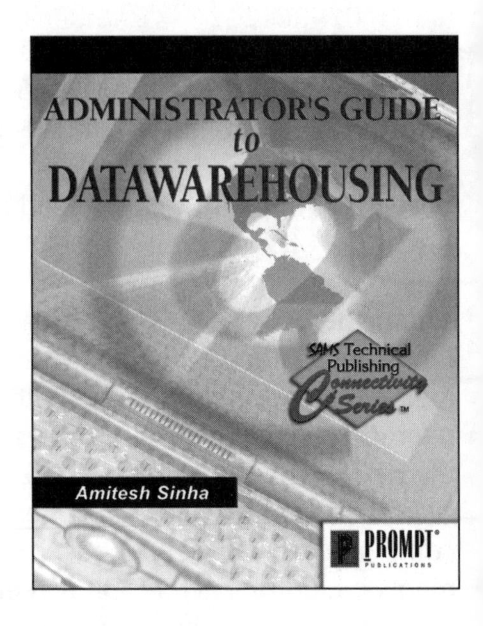

About the book: Datawarehousing is the manipulation of the data collected by your business. This manipulation of data provides your company with the information it needs in a timely manner, in the form it desires. This complex and emerging technology is fully addressed in this book. Author Amitesh Sinha explains datawarehousing in full detail, covering everything from set-up to operation to the definition of terms.

PROMPT® Pointers: Covers On-Line Analytical Processing issues. Addresses set-up of datawarehousing systems. Is designed for the experienced
IT administrator.

Related Titles: *Designing Serial SANS*, ISBN 0790612461, *How the PC Hardware Works*, ISBN 079061250X.

Author Information: Amitesh Sinha has a Masters in Business Administration and over 10 years of experience in the field of Information Technology. Sinha is currently the Director of Projects with GlobalCynex Inc. based in Virginia and has written numerous articles for computer publications.

APPLIED ROBOTICS II

Author: EDWIN WISE
ISBN: 0790612224 • **SAMS#:** 61222
Pages: 304 • **Category:** Projects
Case qty: TBD • **Binding:** Paperback
Price: $29.95 US/$47.95CAN

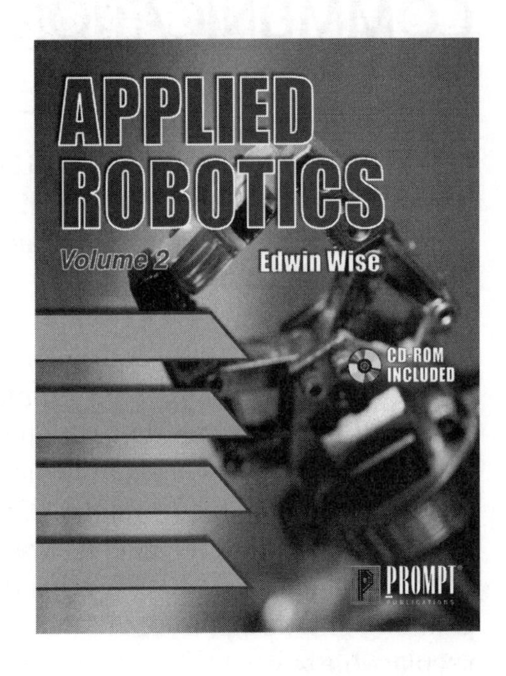

About the book: Edwin Wise builds upon his best-seller, *Applied Robotics* with this book targeted at more advanced hobbyists with development of a larger, more robust, and very practical mobile robot platform. Building on the foundation set in his first text, *Applied Robotics II* has projects to create a larger robot platform suitable for use in the home or outdoors, advanced sensor projects and a great exploration of A1 and control software.

Prompt Pointers: Picks up where *Applied Robotics* left off. Offers an advanced set of projects related to this very hot subject area.

Related Titles: *Applied Robotics*, ISBN 0790611848. *Animatronics*, ISBN 079061294.

Author Information: Edwin Wise is a professional software engineer with twenty years of experience. He currently works in the field of Computer Aided Manufacturing (CAM). His experience includes work on both computer games and educational software. Building robots has been a dream and passion for Edwin for years now. His current project is "Boris," a giant killer robot that can be viewed at http://www.simreal.com/Boris.

GUIDE TO CABLING AND COMMUNICATION WIRING

Author: LOUIS COLUMBUS
ISBN: 0790612038 • **SAMS#:** 61203
Pages: 320 • **Category:** Communications
Case qty: TBD • **Binding:** Paperback
Price: $39.95 US/$63.95CAN

About the book: Part of Sams Connectivity Series, *Guide to Cabing and Communication Wiring* takes the reader through all the necessary information for wiring networks and offices for optimal performance. Columbus goes into LANs (Local Area Networks), WANs (Wide Area Networks), wiring standards and planning and design issues to make this an irreplaceble text.

PROMPT® Pointers:
Features planning and design discussion for network and telecommunications applications. Explores data transmission media. Covers Packet Framed-based data transmission.

Related Titles: *Administrator's Guide to E-Commerce*, by Louis Columbus, ISBN 0790611872. *Exploring LANs for the Small Business and Home Office*, by Louis Columbus, ISBN 0790612291. *Computer Networking for the Small Business and Home Office*, by John Ross, ISBN 0790612216.

Author Information: Louis Columbus has over 15 years of experience working for computer-related companies. He has published 10 books related to computers and has published numerous articles in magazines such as *Desktop Engineering, Selling NT Solutions*, and *Windows NT Solutions*. Louis resides in Orange, CA.

HOW THE PC HARDWARE WORKS

Author: MICHAEL GRAVES
ISBN: 079061250X • **SAMS#:** 61250
Pages: 800 • **Category:** Computer Technology
Case qty: TBD • **Binding:** Paperback
Price: $39.95 US/$63.95CAN

About the book: As the technology surrounding our desktop PCs continues to evolve at a rapid pace, the opportunity to understand, repair and upgrade your PC is attractive. In an era where the PC you bought last year is now "out of date", your opportunity to bring your PC up-to-date rests in this informative text. Renouned author Michael Graves addresses this subject in a one-on-one manner, explaining each category of computer hardware in a complete, concise manner.

Prompt Pointers: Designed to bring a beginner up to a professional level of hardware expertise. Includes new SCSI III implementations, new video standards, and previews of upcoming technologies.

Related Titles: *Exploring Office XP*, ISBN 079061233X, *Designing Serial SANS*, ISBN 0790612461, *Administrator's Guide to Datawarehousing*, ISBN 0790612496.

Author Information: Michael Graves is a Senior Hardware Technician and Network Engineer for Panurgy of Vermont. Graves has taught computer hardware courses on the college level at Champlain College in Burlington, Vermont and The Essex Technical Center in Essex Junction, Vermont. While this is his first full-length book under his own name, his contributions have been included in other works and his technical writing has been the source of several of the more readable user's guides and manuals for different products.

AUTOMOTIVE AUDIO SYSTEMS

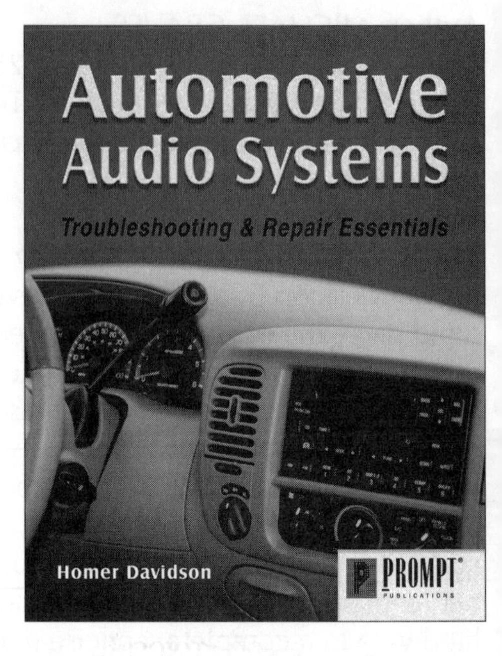

Author: HOMER L. DAVIDSON
ISBN: 0790612356 • **SAMS#:** 61235
Pages: 320 • **Category:** Automotive
Case qty: TBD • **Binding:** Paperback
Price: $39.95 US/$63.95CAN

About the book: High-powered car audio systems are very popular with today's under-30 generation. These top-end systems are merely a component within the vehicle's audio system, much as your stereo receiver is a component of your home audio and theater system. Little has been written about the troubleshooting and repair of these very expensive automotive audio systems. Homer Davidson takes his decades of experience as an electronics repair technician and demonstrates the ins-and-outs of these very high-tech components.

Prompt Pointers: Coverage includes repair of CD, Cassette, Antique car radios and more. All of today's high-end components are covered. Designed for anyone with electronics repair experience.

Related Titles: *Automotive Electrical Systems*, ISBN 0790611422. *Digital Audio Dictionary*, ISBN 0790612011. *Modern Electronics Soldering Techniques*, ISBN 0790611996.

Author Information: Homer L. Davidson worked as an electrician and small appliance technician before entering World War II teaching Radar while in the service. After the war, he owned and operated his own radio and TV repair shop for 38 years. He is the author of more than 43 books for TAB/McGraw-Hill and Prompt Publications. His first magazine article was printed in *Radio Craft* in 1940. Since that time, Davidson has had more than 1000 articles printed in 48 different magazines. He currently is TV Servicing Consultant *for Electronic Servicing & Technology* and Contributing Editor for *Electronic Handbook*.

DESIGNING SERIAL SANS

Author: WILLIAM DAVID SCHWADERER
ISBN: 0790612461 • **SAMS#:** 61246
Pages: 320 • **Category:** Computer Technology
Case qty: TBD • **Binding:** Paperback
Price: $39.95 US/$63.95CAN

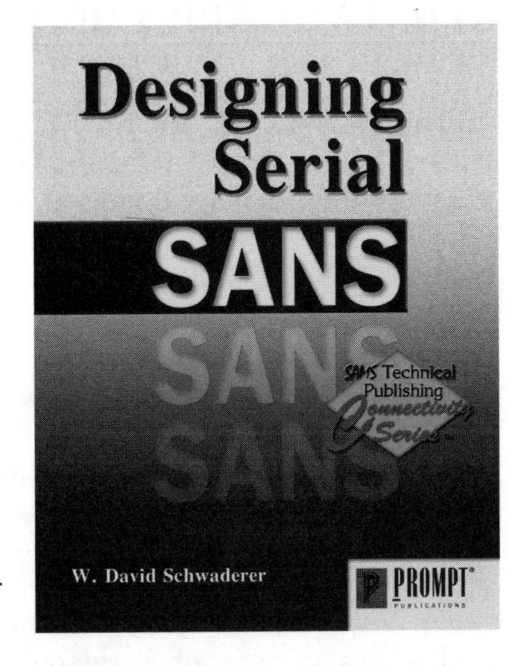

About the book: The use of Serial SANS is an increasingly popular and efficient way to store data in a medium to large corporation setting. Serial SANS effectively stores your company's data away from the traditional server, allowing your valuable server resources to be used for running applications.

Prompt Pointers: Covers Device Specialization Considerations. Explains Media Signals, Data Encoding and Protocols. Discusses SAN hardware building blocks.

Related Titles: *Administrator's Guide to Datawarehousing*, ISBN 0790612496, *How the PC Hardware Works*, ISBN 079061250X.

Author Information: W. David Schwaderer has extensive complex computer system experience and was involved in the creationof two Silicon Valley start-up companies. Schwaderer has a diverse background in connectivity products, personal computer software, and voice DSP based systems. Schwaderer currently resides in Saratoga, CA.

ADMINISTRATOR'S GUIDE TO THE EXTRANET/INTRANET

Author: CONRAD PERSSON
ISBN: 0790612410 • **SAMS#:** 61241
Pages: 304 • **Category:** Computer Technology
Case qty: TBD • **Binding:** Paperback
Price: $34.95 US/$55.95CAN

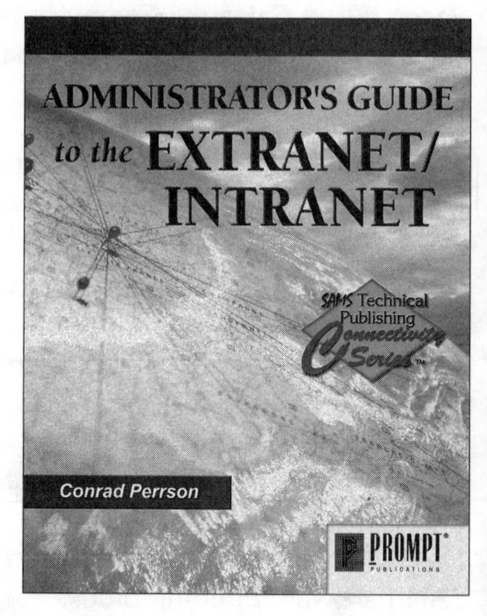

About the book: We are all familiar with the Internet, but few of us have occasion to utilize an Intranet or Extranet application. Both have vast applications related to inner-company communication, customer service, and vendor relations. Both are built similarl to Internet sites, and have many of the same features, issues, and problems. Intranet and Extranet applications are generally under-utilized, even though they provide the opportunity for both communication and financial benefits.

Prompt Pointers: Designed for the Systems Administrator or advanced webmaster. Outlines Intranet/Extranet issues, problems, and opportunities. Discusses hardware and software needs.

Related Titles: *Administrators Guide to E-Commerce*, ISBN 0790611872. *Computer Networking for the Small Business and Home Office*, ISBN 0790612216. *Exploring Microsoft Office XP*, ISBN 079061233X.

Author Information: Conrad Persson is the editor of *ES&T Magazine*, the premier publication for the electronics servicing industry. Conrad has decades of experience related to electronics and computer applications and resides in Shawnee Mission, KS.